T0142881

Springer Earth System Sciences

Series Editors

Philippe Blondel, School of Physics, Claverton Down, University of Bath, Bath, UK

Jorge Rabassa, Laboratorio de Geomorfología y Cuaternario, CADIC-CONICET, Ushuaia, Tierra del Fuego, Argentina

Clive Horwood, White House, Praxis Publishing, Chichester, West Sussex, UK

More information about this series at http://www.springer.com/series/10178

José Herminio Laza

Ichnology of the Lowlands of South America

Paleoichnological Studies in Continental Cenozoic Rocks

 Springer

José Herminio Laza
The Paleontology Division of the La Plata Museum
National University of La Plata
La Plata, Argentina

ISSN 2197-9596   ISSN 2197-960X   (electronic)
Springer Earth System Sciences
ISBN 978-3-030-62599-3   ISBN 978-3-030-62597-9   (eBook)
https://doi.org/10.1007/978-3-030-62597-9

© The Editor(s) (if applicable) and The Author(s), under exclusive license to Springer Nature
Switzerland AG 2020
This work is subject to copyright. All rights are solely and exclusively licensed by the Publisher, whether
the whole or part of the material is concerned, specifically the rights of translation, reprinting, reuse
of illustrations, recitation, broadcasting, reproduction on microfilms or in any other physical way, and
transmission or information storage and retrieval, electronic adaptation, computer software, or by similar
or dissimilar methodology now known or hereafter developed.
The use of general descriptive names, registered names, trademarks, service marks, etc. in this publication
does not imply, even in the absence of a specific statement, that such names are exempt from the relevant
protective laws and regulations and therefore free for general use.
The publisher, the authors and the editors are safe to assume that the advice and information in this book
are believed to be true and accurate at the date of publication. Neither the publisher nor the authors or
the editors give a warranty, expressed or implied, with respect to the material contained herein or for any
errors or omissions that may have been made. The publisher remains neutral with regard to jurisdictional
claims in published maps and institutional affiliations.

This Springer imprint is published by the registered company Springer Nature Switzerland AG
The registered company address is: Gewerbestrasse 11, 6330 Cham, Switzerland

Even the thinnest hair leaves a shadow on the ground

Martín Fierro

A classical book by José Hernández on the history of the "gauchos", the rangers and cowboys of the Argentine Pampas.

Ichnites show the deliberate life activity they represent

José Herminio Laza

Contents

About the Author

José Herminio Laza is a former member of CONICET, the National Research Council of Argentina, forming part of the staff of the Ichnology Division, National Museum of Natural Sciences "Bernardino Rivadavia" and of the Vertebrate Paleontology Division of the La Plata Museum of Natural Sciences, University of La Plata.

Chapter 1
Introduction

Abstract The territorial reconnaissance and scientific study of the Pampas of Argentina by the first biogeographers who visited these regions was that of a very vast region, with unclear borders with the neighboring geographic units. Its main characteristics correspond to a flat landscape with levels of scarce and reduced elevations. Its ample extension develops as an ecotone between the subtropical, warmer and wetter territories in Brazil and northeastern Argentina and the colder and desert-like territories in Patagonia. The Andean Ranges elevation pertaining to the "Quechua" phase that took place during Miocene times provided the clastic and volcanic rocks that covered up the Tertiary "Paranaense" sea deposits and gave the definitive identity to this vast grassland as well as to the hydrographic network that, from the very beginning, had an Atlantic Ocean slope. The lithological composition of the sedimentary sequence presents a high level of homogeneity, being these units formed by slime, sand and clay of a volcanic-pyroclastic origin. These sediments were named as loessoid slimes or silts, or loess-like sediments in several shades of brownish colors, developing in some areas stratification of calcium carbonate in various thicknesses. Most of these strata were later modified by soil forming processes. Such characteristics hinder the reconnaissance of different formational units, partly rectified by researchers, with the reconnaissance of fossil vertebrate faunas corresponding to diverse levels. The development of subsequent study elements to determine the boundaries of different stratigraphic units was based upon the study of paleosols, paleosurfaces and calcareous duricrusts as units apt for regional correlation. Thus, multiple outcrops formed by paleosoil sequences have been thoroughly studied. The development of investigations related to traces and tracks of biotic activity imprinted on these paleosols by various vertebrate and invertebrate animals as well as by plants is even more recent; these studies correspond to the field of paleoichnological research. The continental faunal sequence of the Argentine Pampas, from the Late Miocene to recent times, has been quite useful to build up the chronological framework of the Late Cenozoic of South America. All the type sections of the biostratigraphic units of the Late Cenozoic of this continent are found in the Pampean Region, in some cases overlapping with elements of seven stages/ages. A biozone

© The Author(s), under exclusive license to Springer Nature Switzerland AG 2020
J. H. Laza, *Ichnology of the Lowlands of South America*, Springer Earth System Sciences,
https://doi.org/10.1007/978-3-030-62597-9_1

scheme with biostratigraphic basis for the area has been suggested. The chronostratigraphic units involved correspond to eight stages/ages of this wide stratigraphic
record.

Keywords Pampas plains · Biogeography · Geological sequence · Paleontology

1.1 Introduction

Ichnology has recently gained the field of continental territories thanks to the application of concepts resulting from the recognition of ichnofacies as a way of interpreting
the sets of traces according to the sites and environments where they occur. In the
present work, the record spans the Late Cenozoic of the Pampean region (from
the Late Miocene to the Holocene; Fig. 1.1), while listing and describing various
ichnotaxons, providing their geographic location and biostratigraphic position.

Paleosols are recognized as important units bearing fossil traces, their specific
features and their bio-chronological importance, due to their relative short periods
of persistency in the geological record.

The insect ichnites related to current taxons are supported by zoological–environmental knowledge and their comparison with other paleontological studies carried
out in the region, referring to vertebrates and paleobotany, as well as the pertaining
geological observations.

1.2 The Geological Context

1.2.1 The Territory

The Chaco-Pampa plains outstand as a very large geomorphological unit in South
America, of somewhat uncertain borders in relation to the neighboring units
(Fig. 1.2). Such plain is defined as occurring at less than 200 m altitude above sea level,
stretching from Bolivia and Paraguay toward the South as far as the Colorado River
in Argentina and the corresponding Atlantic Ocean coast, including the Argentine
provinces of Buenos Aires, Southern Córdoba, Santa Fe, Entre Ríos and Corrientes,
the whole Uruguayan territory and the Brazilian state of Rio Grande do Sul. This
region is defined toward the West by the Sub-Andean and Pampean Hills, and it
stretches toward the East enclosing the hilly, bedrock Precambrian and Paleozoic
cratonic systems known as Tandilia and Ventania in Buenos Aires province. Martín
De Moussy (1873) named as "Pampasia" the southern part of this huge system of
plains, coinciding partly with the criteria of some biogeographers such as Ringuelet
(1961), Cabrera and Willink (1973) and Morrone (2001).

"Pampasia" constitutes a vast ecotone developed between the warmer and wetter
Brazilian lands and those colder and more arid domains in northern Patagonia.

Fig. 1.1 Chronologic
scheme of the Late Cenozoic
of South America Modified
from Cione et al. 2015

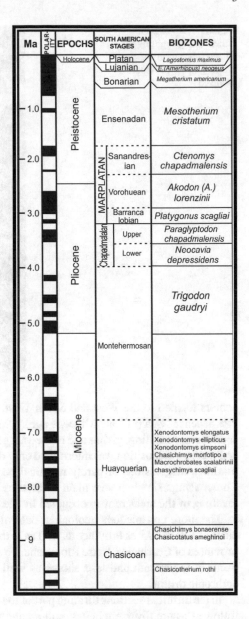

1.2.2 The Sedimentary Units

By the Middle Miocene, the final uplifting of the Andean Ranges started (a stage known as the "Quechua" phase), being the drainage system outlined increasingly more consistently, particularly that of Atlantic Ocean slope. These hydrological systems carried the sediment waste through vast continental areas and their selection took place as descending toward the East, as they drifted away from the deposition

Fig. 1.2 Map of South
America, highlighting the
Chaco sub-region and the
biogeographic provinces
involved

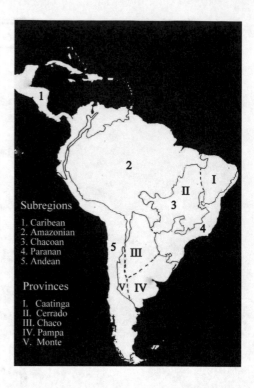

Subregions

1. Caribean
2. Amazonian
3. Chacoan
4. Paranan
5. Andean

Provinces

I. Caatinga
II. Cerrado
III. Chaco
IV. Pampa
V. Monte

centers located to the West and South West. Gradually, the basin occupied by the
"Paranaense" Sea (of Late Miocene age), which was in full regression, began to get
sedimentary infilling, giving rise to the vast plain system that exists today. The thick
sedimentary deposition was interrupted only during the Pleistocene and Holocene by
marine transgressions of barely marginal penetration in the continental territories.
Thus, "Pampasia" is a vast plain created by the sedimentary infilling and surface
levelling of the areas mostly occupied by the "Paranaense" Sea (Fig. 1.3).

The more conspicuous geological outcrops of the region correspond to the sea
cliffs of Buenos Aires Province as well as to the Paraná River gullies in the Argentine
provinces of Corrientes, Entre Ríos, Santa Fe and Buenos Aires. Smaller expositions
occur in the stream and lake shores as well as in depressions and excavations of
anthropic origin.

In the different sections forming part of the sequence, several paleo-edaphic levels
followed where diverse ichnites, and among them, nests of numerous insects, have
been thoroughly recorded.

The continental sedimentary sequence of the Pampean area spans from the Cretaceous to the Recent (Russo et al. 1979), whereas the outcropping succession begins
in the Late Miocene and spans until nowadays. Such sequence is characterized by
a significant sedimentary and chromatic homogeneity (Teruggi et al. 1957), as well
as by numerous lateral surface changes that hinder the reconnaissance of boundaries
among formations. These difficulties were overcome partly by several generations

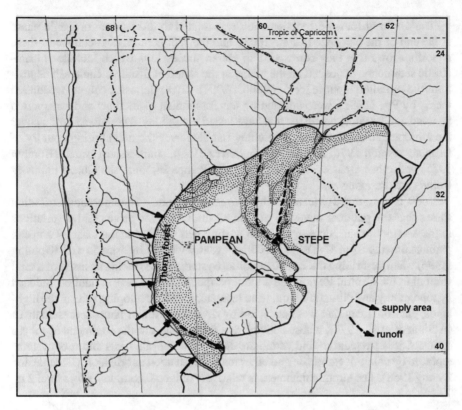

Fig. 1.3 Map of the Pampean Province, with the indication of zones of sedimentary supply and drainage (drawing by Marcela Tomeo)

of researchers that used the paleontological content of different levels as a diagnostic element of subdivision (D´Orbigny 1841; Darwin 1846; Bravard 1857a, b; Ameghino 1876, 1907, 1908, 1910; Kraglievich 1934, 1952, 1959; Cione and Tonni 1995a, b, c). The use of criteria and concepts oriented to establish more accurate stratigraphic instruments developed the study of paleosols, paleosurfaces and calcareous crusts as possible correlation tools (Teruggi et al. 1974; Zárate and Fasano 1984).

The "Pampean sediments" (Fidalgo et al. 1975) are mostly formed by subordinate silt, sand and clay of volcanic and pyroclastic origin, whereas the sediments of metamorphic origin are poorly represented (Teruggi et al. 1957). These sediments are called loess or loessoid materials; they present massive aspect and occasionally a faint stratification, while their color comprises a wide shade of browns, with intercalations ranging from yellowish to greenish tones. In some sectors, the precipitation of calcium carbonate (locally named as "tosca") is in the shape of strata, with thicknesses ranging from a few centimeters to several meters. The "tosca" also shows "doll" shapes or root-like aspects.

The origin of loessoid sedimentation was related to the Andean orogenic phase at the end of the Miocene (Zárate 2003), and the sedimentological control of these deposits shows that they correspond to eolian successions which include volcani- clastic sediments, deposited on the plains in the shape of distal volcanic ash deposi- tion (the so-called tephrite loess) (Bellosi 2004). These deposits, called "weathered loess" by Pye (1987), were affected by the Patagonian glaciations and transported by wind and streams, processes that have lasted until today. These loess strata cover the aforementioned territories, stretching out also over the neighboring country of Uruguay (Antón 1976; Panario and Gutiérrez 1999), southeastern Brazil (Bombín 1975) and other areas even to the north of "Pampasia," like the Chaco plains in Bolivia and Paraguay.

Loess is recognized as an allochthonous sediment, and its geographic distribution shows that the parental materials present textural changes produced by variations in the energy of transport agents. Their sorting was verified in the Chaco-Pampean plain sediments from W-SW to E-NE, an event reviewed by Iriondo and Krohling (1996) when describing the Pampean Eolian System. This system consisted of a sand stratum in the Central Region and a loess peripheral belt. These eolianites, induced by topography and climatic action, were later stabilized, developing soils in the high- lands, environments that were occupied by diverse flora and fauna representatives (Ahlbrandt et al. 1978), whose activity was then printed in the sediments. In these deposits, it is frequent to find vertebrate caves, invertebrate nests and root systems replaced by calcium carbonate and other materials, named as "dikaka" (Glennie and Evamy 1968). The stratum thickness is relatively reduced, commonly less than 2 m.

1.2.3 Paleosols

The loess strata frequently present evidence of paleosoil formation. These paleosols are recognized by their profiles, formed by horizons as well as by the included biological evidence (crotovines, calcium carbonate root molds, insect nests) and in the laboratory by their microscopic features. The soils and consequently the pale- osols are not mere deposits but the result of the postdepositional modification of alluvial, marsh, lacustrine and eolian accumulations on which pedogenesis acted, modifying the surface of the exposed rocks, producing different degrees of weath- ering. Pedogenesis depends upon the frequency of depositional events, distance of the initial sources of the materials and their type, position and fluctuation of the hydrological profile, temperature and precipitation (Hasiotis 2000). The resulting kind of soils develops like a microcosmos, with their own physical, chemical and biological distinctiveness. The existence of periods of landscape stability during the Late Cenozoic in "Pampasia" allowed the formation of paleosols in different paleo- environments. Frenguelli (1926) is the first in our country mentioning observations related to paleosols, pointing out rodent caves and insect nests, proposing the term root casts to the accumulations of calcium carbonate surrounding the rootlets, all of them unequivocal signs of paleo-edaphic levels. Subsequent stratigraphic and sedimentological studies of the Pampean Formation in the cliffs between the cities of Mar del Plata and Miramar, Buenos Aires province (Kraglievich 1952; Teruggi et al. 1957) mentioned the presence of paleosols. Teruggi and Imbellone (1986) listed the diverse contributions made about paleosols of the Pampean Region until that moment, and those studies by Fidalgo et al. (1973a, b), Zárate and Fasano (1984); Zárate et al. (2002), Zinck and Sayago (2001), Iriondo and García (1993), Kröhling (1999a) and others mentioned in the bibliography.

Research carried out in the Pampean plain and the pre-Cordilleran valleys (Sayago et al. 2001) indicate the alternation of pedogenic cyclic nature—stratum cyclic nature—a phenomenon globally observed and exposed in the so-called K cycles by Butler (1959). Allen and Goss (1973) stated that most of the paleosols developed on the loessoid materials are truncated and the lack of an A horizon is the result of erosion after pedogenesis, with possible clay translocation to calcareous lower horizons. Instead, Teruggi and Imbellone (1986) claimed that there is a close relationship among deposits of loess stratum, pedogenesis and truncation, phenomena repeated cyclically. They proposed a hypothesis that the overlapped paleosols, without apparent illuvial horizons, were not truncated in all cases, but are instead complete. The A horizons would be altered by total oxidation of the organic matter and the development of the overlying paleopedological profile, whose base, due to the small stratum thickness, modifies the properties of the underlying paleosol with which its B horizon acts as the horizon C of the underlying bed. The cyclic sequence of overlapping paleosols is frequent in the Pampasia and nearby zones. Teruggi et al. (1974) studied one of these sequences in the Mar del Plata-Miramar cliffs consisting of twelve paleosols. Imbellone and Teruggi (1986) described six overlapping paleosols in excavations near Mar del Plata, 7.85 m thick, that they identified as mollisols or alfisols. During the excavations for the building of the Teatro Argentino in the city of La Plata, Riggi et al. (1986) recognized ten levels with soil remnants of up to 28 m thick. Besides, Sayago et al. (2001) mentioned a sequence of 28 paleosols and 40 m of thickness for the intermountain valleys of the Tucumán Precordillera (the Tafí del Valle Formation). Despite all these works, the studied and named paleosols are scarce (Rabassa et al. 2005).

Besides, the paleosols are worthy of study in the laboratory. Some of these, carried out through thin cuts, allow observing the micromorphology of their structures, facilitating the identification of numerous biological phenomena developed in them. The influence of the animal activity is visible by the study with microscope of thin soil sections, enabling the recognition of several microstructures directly associated with such activity: fecal and inorganic pellets of construction, excavation, coverings and walls. The recognition of some of these signs allows the identification of various animal groups.

The contribution of the soil animals to production and maintenance of their factory is a vital part in their function. The concept of ichnofactory was introduced by Ekdale and Bromley (1983) and corresponds to "those aspects of the sediment internal texture and structure that result from the bioturbation and bioerosion at all scales." Thus, the studies to the ichnofactory have become an important tool for ichnological analysis, allowing documenting and comparing the existence of bioturbation and relative chronology of the in-faunal association. For this purpose, several comparative scales of the development degree of the ichnofactory in paleosols, one of them, proposed by Reineck (1963) and adopted by Taylor and Goldring (1993), consist of seven degrees to recognize the perturbation density. According to Ekdale and Bromley (1983), it is possible to consider the soil structure on the whole as an ichnofactory, for being the result of direct or indirect bioturbations.

A relevant fact in the study of an ichnofactory is to recognize the patterns of the organism (taxon) linked to the substratum and its modification throughout time in relation to environmental changes. Bown and Kraus (1983) suggested a method based on the detailed record and frequency comparison of particular fossil traces in different types of paleosols and horizons. In turn, Genise et al. (2004) provided a complete overview of the study of ichnofactories from different sites, proposing a ternary diagram for the index evaluation of bioturbation, ichnofactory and pedofactory, while defining that the factory produced by roots and animal traces should be considered as ichnofactory, whereas the factory produced by other soil characters, of physical or chemical nature, should be called pedofactory.

The animal role in the formation of soil microstructures can be divided in three basic categories: (a) living matter death and disintegration; (b) formation of zoogenic microstructures in the soil matrix; and (c) excavation and construction activity (Rusek (1985).

The stratigraphic scheme suggested by Fidalgo et al. (1973b) for the Late Pleistocene–Holocene recognized and nominated several paleosols: In the upper third of the Bonaerian stratigraphic unit, these authors identified the "suelo sin nombre" (i.e., "unnamed soil"); likewise, at the top of the Guerrero Member of the Luján Formation, the "Puesto Callejón Viejo Soil" was recognized at the Pleistocene–Holocene boundary, and the "Puesto Berrondo Soil" was found in the Holocene. Besides, Nabel (1993) and Nabel et al. (1997), during sedimentological and magnetostratigraphic studies in several sites in northern Buenos Aires province, identified and nominated different paleosols. One of them, of relevant magnetostratigraphic importance served as the boundary between the Brunhes–Matuyama chrons: the "El Tala Geosoil," developed on the Ensenada Formation. Above the "El Tala Geosoil," the "Hisisa Geosoil" was developed, of very wide areal extent.

The recurrence of paleosols provides significant information about the dynamic processes of change, of local and regional nature. By observing them, Bown and Kraus (1987) and Retallack (1988) suggested the use of scales to determine the degree of development of the soil horizons.

1.2.4 Chronology

The continental faunal sequence from the Late Miocene to Recent of the Argentine "Pampasia" area provided the basic scheme for building up the chronological scale of the South American continent—though it is arguable its correlation with the rest of the continent (Cione and Tonni 1999; Cione et al. 2015). All type sections of different biostratigraphic units of the Late Cenozoic are found in Argentine territory and, more accurately, in the Pampa Region (Cione and Tonni 1995a, b, c, 1996; Cione et al. 2015) where sediments of seven different ages appear overlapped, sometimes continuously. The area comprises plains stretching as far as Uruguay and southern Brazil, where their sediments have been correlated (Mones 1979; Ubilla and Perea 1999; Panario and Gutiérrez 1999; Perea and Martínez 2004; Oliveira 1999; Martínez and Ubilla 2004).

Cione and Tonni (1999) suggested a scheme considering biozones as biostratigraphic basis for the Pampean area. Later, further research added new biozones to the initial scheme (Cione et al. 2015). Such biozones constitute the following chronostratigraphic units:

- Chasicoan: biozones of Chasicotherium rothi (Tonni et al. 1998), Chasicotatus ameghinoi (Tonni et al. 1998) and Chasichimys bonaerense (Verzi et al. 2008).
- Huayquerian: (ca. 9.0–6.8 Ma; Woodburne et al. 2006). Biozones of Macrochorobates scalabrinii (Tonni et al. 1998), Chasichimys scagliai, Chasichimys morphotype a, Xenodontomys simpsoni, Xenodontomys ellipticus and Xenodontomys elongatus (Verzi et al. 2008).
- Montehermosan: (ca. 6.8–5.0 Ma; Woodburne et al. 2006). Biozone of Trigodon gaudryi and Neocavia depressidens (Cione and Tonni 2005) = Eumysops lacviplicatus (Tomassini and Montalvo 2013; Tomassini et al. 2013).
- Chapadmalalan: (ca. 5.0–3.3 Ma; Woodburne et al. 2006). Biozone of Paraglyptodon chapadmalensis.
- Marplatan: (ca. 3.2–2.0 Ma; Woodburne et al. 2006).
- Barrancalobian: (around 3.2–3.0 Ma; Woodburne et al. 2006). Biozone of Platygonus scagliai (Cione and Tonni 1995a).
- Vorohuan: (around 3.0–2.4 Ma; Woodburne et al. 2006). Biozone of Akodon (A) lorenzinii (Cione and Tonni 1995a).
- Sanandresian: (around 2.4–2.0 Ma; Woodburne et al. 2006) Biozone of Ctenomys chapadmalensis (Cione and Tonni 2005).
- Ensenadan: (ca. 2.0–0.4 Ma; Woodburne et al. 2006). Biozone of Mesotherium cristatum (Cione and Tonni 2005).
- Bonaerian: (ca. 0.4–0.125 Ma; Woodburne et al. 2006). Biozones of Ctenomys kraglievichi (Verzi et al. 2004) and Megatherium americanum (Cione and Tonni 1999).
- Lujanian: (ca. 0.125–0.08 Ma; Woodburne et al. 2006). Biozone of Equus (Amerhippus) neogaeus (Cione and Tonni 1999).

- Platan: (ca. 0.08 Ma. to the Late Holocene). Biozone of Lagostomus maximus (Cione and Tonni 1999). In the present work, the mentioned bio- and chronostratigraphic scheme is adopted.

References

Ahlbrandt T, Andrews S, Gwynne D (1978) Bioturbation in eolian deposits. J Sediment Petrol 48(3):839–848
Allen B, Goss D (1973) Micromorphology of paleosols from the semi-arid plains of Texas. In: Rutheford G (ed) Soil microscopy. The Limestone Press, Canada. pp 511–525
Ameghino F (1876) Ensayo de un estudio de los terrenos de transporte cuaternarios de la Provincia de Buenos Aires. Obras Completas y Correspondencia Científica II:53–137
Ameghino F (1907) Enumeración cronológica y crítica de las noticias sobre las tierras cocidas y las escorias antrópicas de los terrenos sedimentarios neógenos de la Argentina aparecidas hasta fines del año 1907. Obras Completas y Correspondencia Científica XVIII:69–269
Ameghino F (1908) Las formaciones sedimentarias de la región litoral de Mar del Plata y Chapadmalán. Anales Museo Nac de Historia Nat 3(10):343–428
Ameghino F (1910) La actividad geológica del yacimiento antropolítico de Monte Hermoso. Congreso Científico Internacional Americano, Actas, pp 1–6
Antón D (1976) Relaciones entre las superficies de aplanamiento y los suelos en el Uruguay. UDIA, INTA, Suplemento 33:409–413
Bellosi E (2004) Sedimentological control the Coprimisphaera Ichnofacies. In: First International Congress of Ichnology. Abstract book: 19. Trelew, Argentina
Bombín M (1975) Afinidade paleoecológica, cronológica e estratigráfica do componente de Mega-mamíferos na biota do Cuaternario terminal da Provincia de Buenos Aires, Uruguay e Río Grande do sul (Brasil). Comunicaçoes do Mus. Cs. da PUCRGS, pp 1–28
Bown T, Kraus M (1983) Ichnofossils of the alluvial Willwood Formation (Lower Eocene), Bighorn Basin, Northwestern Wyoming, U.S.A. Palaeogeogr Palaeoclimatol Palaeoecol 43:95–128
Bown T, Kraus M (1987) Integration of channel and floodplain suites. I. Developmental sequence and lateral relations of alluvial paleosols. J Sediment Petrol 57:587–601
Bravard A (1857a) Observaciones geológicas sobre diferentes terrenos de transporte en la Hoya del Plata. Biblioteca del diario La Prensa, Buenos Aires
Bravard A (1857b) Estado físico del territorio. Geología de las pampas. In: Registro Estadístico del Estado de Buenos Aires, pp 1–22
Butler B (1959) Periodic phenomena in landscapes, a basis for soil studies. CSIRO, Australia, Soil Publications 14:4–20
Cabrera A, Willink A (1973) Biogeografía de América Latina. Monografía N° 13. Serie Biología. OEA, Washington, D.C
Cione A, Tonni E (1995a) Chronostratigraphy and "Land-mammal ages" in the Cenozoic of Southern South America: principles, practices, and the "Uquian" problem. J Paleontol 69:135–159
Cione A, Tonni E (1995b) Bioestratigrafía y cronología del Cenozoico superior de la Región pampeana (47–74). In: Evolución biológica y climática de la región pampeana durante los últimos cinco millones de años. In: Alberdi M, Leone G, Tonni E (eds) Monografía N° 12, Museo Nacional de Ciencias Naturales, Madrid. p 423
Cione A, Tonni E (1995c) Los estratotipos de los pisos Montehermosense y Chapadmalalense (Plioceno) del esquema cronológico sudamericano. Ameghiniana 32:369–374
Cione A, Tonni E (1996) Inchasi, a Chapadmalalan (Pliocene) locality in Bolivia. Comments on the Pliocene-Pleistocene continental scale of southern South America. J S Am Earth Sci 9:221–236

Cione A, Tonni E (1999) Biostratigraphy and Chronological scale of uppermost Cenozoic in the Pampean area, Argentina (23–52). In: Rabassa J, Salemme M (eds) Quaternary vertebrate paleontology in South America. Special volume of Quaternary of South America and Antarctic Peninsula, A.A. Balkema Publishers, Rotterdam, pp 12, 310

Cione A, Tonni E (2005) Bioestratigrafía basada en mamíferos del Cenozoico de la Provincia de Buenos Aires, Argentina. XVI Congreso Geológico Argentino, Relatorio, pp 183–200

Cione AL, Gasparini GM, Soibelzon E, Soibelzon L, Tonni EP (eds) (2015) The great American biotic interchange. A South American perspective. Springer Briefs in Earth system sciences. Springer

Darwin C (1846) Geological observations on South America. Being the third part of the Geology of the voyage of the Beagle, under the command of Cap. Fitzroy, R.N. during the years 1832 to 1836. London

De Moussy M (1873) Description geographique et statistique de la Confederation Argentine. 2ª Ed. Paris

D´Orbigny A (1841) Voyage dans l´Amérique méridionale. III, 3° partie: Géologie; 4° partie: Paléontologie. Paris

Ekdale A, Bromley R (1983) Trace fossils and ichnofabrica in the Kjolby Gaard Marl, Upper Cretaceous, Denmark. Bull Geol Soc Denmark 31:107–119

Fidalgo F, Colado U, De Francesco F (1973a) Sobre las ingresiones marinas cuaternarias en los partidos de Castelli, Chascomús y Magdalena (provincia de Buenos Aires). V Congreso Geológico Argentino. Actas III: pp 227–240

Fidalgo F, De Francesco F, Colado U (1973b) Geología superficial en las Hojas Castelli, J. M. Cobo y Monasterio (provincia de Buenos Aires). V Congreso Geológico Argentino. Actas IV: pp 27–39

Fidalgo F, De Francesco F, Pascual R (1975) Geología superficial de la llanura bonaerense. In: Relatorio del VI Congreso Geológico Argentino, pp 103–138

Frenguelli J (1926) Sulle concrezioni calcaree intorno alle radicidi vegetali viventi. Bolletino della Societá Geol ital 45(1):85–90

Genise J, Bellosi E, González M (2004) An approach to the description and interpretation of ichnofabrics in paleosols. In: Mc Ilroy (ed) The application of ichnology to palaeoenvironmental and stratigraphic analysis. Geological Society. Special Publications, London, pp 228: 355–382

Glennie K, Evamy B (1968) Dikaka: Plants and plant-root structures associated with aeolian sand. Palaeogeogr Palaeoclimatol Palaeoecol 23:77–87

Hasiotis S (2000) The invertebrate invasion and evolution of Mesozoic soil ecosistems: the ichnofossil record of ecological innovations. Paleontological Soc Pap 6:141–169

Imbellone P, Teruggi M (1986) Morfología y micromorfología de toscas de algunos paleosuelos en el área de La Plata. Cienc del Suelo 4:209–215

Iriondo M, García M (1993) Climatic variations in the Argentine plains during the last 18,000 years. Palaeogeogr Palaeoclimatol Palaeoecol 1010:209–220

Iriondo M, Krohling D (1996) Los sedimentos eólicos del noreste de la llanura pampeana (Cuaternario superior). XIII Congreso Geológico Argentino and III Congreso de exploración de hidrocarburos, vol 4, Actas, Buenos Aires, pp 27–48

Kraglievich L (1934) La antigüedad pliocena de las faunas de Monte Hermoso y Chapadmalal, deducidas de su comparación con las que le precedieron y sucedieron. Imprenta El Siglo Ilustrado, Buenos Aires

Kraglievich JL (1952) El perfil geológico de Chapadmalal y Miramar, provincia de Buenos Aires. Rev Museo Municipal Ciencias Naturales y Tradicionales, Mar del Plata 1(1):8–73

Kraglievich JL (1959) Contribución al conocimiento de la geología cuartaria en la Argentina. Museo argentino ciencias naturales "Bernardino Rivadavia". Comunicaciones 1:1–19

Kröhling D(1999a) Upper quaternary geology of the lower Carcarañá Basin, North Pampa Argentina. Quatern Int. 57–58:135–148

Martínez S, Ubilla M (2004) El Cuaternario en Uruguay. In: Bossi L (eds) Geología del uruguay, Montevideo, pp 195–227

Mones A (1979) Terciario del Uruguay. Revista de la Facultad de Humanidades y Ciencias. Serie Ciencias de la Tierra 1(1):1–27

Morrone J (2001) Biogeografía de América Latina y el Caribe. SEA-CYTED, Mexico and Spain, p 144

Nabel P (1993) The Brunhes-Matuyama boundary in Pleistocene sediments of Buenos Aires Province, Argentina. Quatern Int 17:79–85

Nabel P, Etchichury M, Tófalo R (1997) Estratotipo del límite superior de la Formación Ensenada: Geosuelo El Tala. Primer taller sobre sedimentología y medio ambiente. In: Asociación Argentina de Sedimentología. Resúmenes, Buenos Aires, pp 11–12

Oliveira E (1999) 5. Quaternary vertebrates and climates of southern Brazil. In: Rabassa J, Salemme M (eds) Special volume of quaternary of South America and Antarctic Peninsula, 12. A.A. Balkema Publishers, Rotterdam, pp 61–73

Panario D, Gutiérrez O (1999) The continental uruguayan Cenozoic: an overview. Quatern Int 62:75–84

Perea D, Martínez S (2004) Estratigrafía del Mioceno-Pleistoceno en el litoral suroeste de Uruguay. In: Bossi L (ed) Geología del uruguay. Montevideo, pp 105–124

Pye K (1987) Aeolian dust and dust deposits. Academic Press, New York, p 334

Rabassa J, Coronato A, Salemme M (2005) Chronology of the Late Cenozoic Patagonian glaciations and their correlation with biostratigraphic units of the Pampean region (Argentina). J S Am Earth Sci 20:81–103

Reineck H (1963) Sediment refuge im Bereich der sudlichen Nordsee. Abh der seckenbergische naturfoschende Gesellschaft 505:1–138

Retallack G (1988) Field recognition of paleosols. Geol Soc Am. Special Paper 216, p 20

Riggi J, Fidalgo F, Martínez O, Porro N (1986) Geología de los "Sedimentos Pampeanos" en el Partido de La Plata. Rev Asoc Geol Argent 41:316–333

Ringuelet R (1961) Rasgos fundamentales de la zoogeografía de la Argentina. Physis 22:151–170

Rusek J (1985) Soil microestructures—contributions on specific soil organisms. Quaestiones Entomologicae 21:497–514

Russo A, Ferello R, Chebli G (1979) Llanura Chaco—Pampeana. Geología Regional Argentina. Academia Nacional de Ciencias, Córdoba, pp 139–184

Sayago J, Collantes M, Karlson A, Sanabria J (2001) Genesis and distribution of the late pleistocene and holocene loess of Argentina: a regional approximation. Quatern Int 76:247–257

Taylor A, Goldring R (1993) Description and analysis of bioturbation and ichnofabric. J Geol Soc London 150:141–148

Teruggi M, Etchichury M, Remiro J (1957) Estudio sedimentológico de los terrenos de las barrancas de la zona de Mar del Plata—Miramar. Revista Museo Argentino Ciencias Naturales "Bernardino Rivadavia". Geología, vol 4. Buenos Aires, pp 107–250

Teruggi M, Andreis R, Mazzoni M, Dalla Salda L, Spalletti L (1974) Nuevos criterios para la estratigrafía del Cuaternario de las Barrancas de Mar del Plata-Miramar. LEMIT. Anales. Serie II 268:135–148

Teruggi M, Imbellone P (1986) Paleosuelos de la región pampeana. Segunda Jornada de suelos de la región pampeana. La Plata. Actas, pp 40–66

Tomassini RL, Montalvo C (2013) Taphonomic modes on fluvial deposits of the Monte Hermoso-Formation (early Pliocene), Buenos Aires province. Argentina. Palaeogeogr. Palaeoclimatol. Palaeoecol. 369:282–294

Tomassini RL, Montalvo CI, Deschamps CM, Manera T (2013) Biostratigraphy and biochronolo-gyof the Monte Hermoso Formation (early Pliocene) at its type locality, Buenos Aires province, Argentina. Journal South American Earth Sciences 48:31–42

Tonni E, Pardiñas U, Verzi D, Noriega J, Scaglia O, Dondas A (1998) Microvertebrados pleistocénicos del sudeste de la provincia de Buenos Aires (Argentina): bioestratigrafía y paleoambientes. V Jornadas geológicas y geofísicas bonaerenses. Actas I:73–83

Ubilla M, Perea D (1999) 6. Quaternary vertebrates of Uruguay. A biostratigraphic, biogeographic and climatic overview. In: Rabassa J, Salemme M (eds) Quaternary vertebrate paleontology in

South America. Special volume of Quaternary of South America and Antarctic Peninsula, 12, A.A. Balkema Publishers, Rotterdam, pp 75–90

Verzi D, Deschamps C, Tonni E (2004) Biostratigraphic and paleoclimatic meaning of the Middle Pleistocene South American rodent Ctenomys kraglievichi (Caviomorpha, Octodontidae). Palaeogeogr Palaeoclimatol Palaeoecol 212:315–329

Verzi D, Montalvo C, Deschamps C (2008) Biostratigraphy and biochronology of the Late Miocene of central Argentina: evidence from rodents and taphonomy. Geobios 41:145–155

Woodburne M, Cione A, Tonni E (2006) Central American provincialism and the Great American Biotic Interchange. Publicación Especial del Instituto de Geol y Centro de Geocienc de la Univ Nac Autónoma de México 4:73–101

Zárate M, Fasano J (1984) Características de la sedimentación pleistocena en la zona de Chapadmalal, provincia de Buenos Aires. IX Congreso Geológico Argentino, Actas IV. San Carlos de Bariloche, Río Negro, Argentina, pp 57–75

Zárate M, Kemp R, Blasi A (2002) Identification and differentiation of Pleistocene paleosols in the northern pampas of buenos aires, argentina. J S Am Earth Sci 15:303–313

Zárate M (2003) The loess record of Southern South America. Quat Sci Rev 22:1987–2006

Zinck J, Sayago J (2001) Loess-palesol sequence of La Mesada in Tucumán province, northwest Argentina. Characterization and paleoenvironmental interpretation. J S Am Earth Sci 12(3):293–310

Chapter 2
The Paleontological Context. Ichnology

Abstract Except for some previous records, the first attempts for the study of the lying tracks on sediments started at the beginning of the nineteenth century, as well as the efforts for creating a suitable terminology. Its installation as a scientific discipline began at the end of the nineteenth century. By middle twentieth century, a modern and organized scheme for the classification of ichnofossils and their location in ichnofacies was proposed. Firstly, the only recognized ichnofacies corresponded to marine environments and were defined by bathymetric levels and particular ichnofossils. Immediately, ichnofacies corresponding to transition environments from those marine to continental were identified. In 1985, the ichnological nomenclature became part of the International Code of Zoological Nomenclature. During the 1970s and 1980s, there were papers describing ichnites of continental origin, events that quickly led to the organization of the formal reconnaissance of environments carrying continental tracks, giving rise to successive ichnofacies. Apart from the paleosurfaces, the ichnological content is rich and quite varied in paleosols and so are the tracks of insect nests, solitary or in groups in these paleosols.

Keywords Ichnology · Continental ichnofacies · Ichnites in paleosols

2.1 Scheme of Continental Ichnofacies

Although inert in themselves, the constructions made by a living organism contain a true flow of information about the needs and intentions of its builder (Lovelock 1979).

Leaving aside some reference to coprolites toward the end of the seventeenth century, the beginnings of ichnology go back to the beginning of the nineteenth century with the description of vertebrate footprints and traces from marine rocks; additional tasks slowly developed and Ichnology attained status of scientific discipline in the last decades of the nineteenth century (Osgood 1975). The term Ichnology was proposed by Buckland in 1836, during the study of invertebrate trace fossils which were interpreted as belonging to algae ("fucoides") until Nathorst (1881), using the Uniformitarism Principle, proved their zoological origin.

© The Author(s), under exclusive license to Springer Nature Switzerland AG 2020   15
J. H. Laza, *Ichnology of the Lowlands of South America*, Springer Earth System Sciences,
https://doi.org/10.1007/978-3-030-62597-9_2

The attempts for regulating the ichnotaxonomy began very early: In 1853, Jardine proposed naming the tracks with a name finished in ichnus in order to differentiate them from the corporal fossil names, a proposal which was then supported by Seilacher (1953) and Häntzschel (1962).

At the beginning of the twentieth century, Abel and then his disciples continued the ichnological studies introducing main concepts such as bioturbation and biostratification. It was Abel (1935) who also revitalized the vertebrate paleoichnology in Europe, whereas Gilmore and Peabody in North America (from the 1920s to the 1940s) and Rodolfo Casamiquela in South America did so during the 1960s. Richter (1927) began the systematic study which continued until the 1950s when Seilacher (1953) proposed a new methodology in the study and classification of ichnological materials, outlining the model of ichnofacies. He distinguished several marine ichnofacies, developed in specific environments at different bathymetric levels and composed of particular ichnofossils (Seilacher 1964, 1967). The Seilacherian proposal not only assigned the traces to a paleontological nature but also considered the biological activity and sedimentary processes. The initial scheme was improved, being integrated with eight marine ichnofacies and a continental one called Scoyenia, which corresponded to "sandstone and non-marine lutites, commonly red strata" (Seilacher 1953) formed by traces of marine and continental environments of salty environments. The Scoyenia ichnofacies reflects the superposition of sets of fossil traces associated with changes in the substratum consistency, water saturation, generally associated with changes of water levels. Scoyenia shows low ichnodiversity, with the presence of meniscated holes, systems of bilobed holes and arthropod trails. The works by Häntzschel (1962, 1965) accelerated the development of the discipline, marked by a continuous refinement (Osgood 1975). In these works, the information from the stratigraphic record and that from the study of current environments cohabited interactively.

Once the Scoyenia ichnofacies was established, the study of continental ichnology detained (Seilacher 1964). Frey et al. (1984a) questioned the boundaries of such ichnofacies, for being extremely broad and varied, arguing that Scoyenia was only one of the continental ichnofacies which was poorly known and indefinite.

In spite of the early attempts for regulating the norms of ichnologic nomenclature, these did not have official status until the publication of the third edition of the International Code of Zoological Nomenclature in 1985. Until that moment, the continental traces were associated with footprints of vertebrates, coprolites, nests, bones with marks and traces and on plants, whereas those of terrestrial invertebrates were considered doubtful and, incorrectly and scornfully called "snail holes" and "tracks and trails" (Hasiotis and Bown 1992). The 1970s and 1980s showed contributions about continental ichnites, in fluvial and lacustrine environments, interpreted ecologically and ethologically (Bromley and Asgaard 1979). Ratcliffe and Fagerstrom (1980) listed diverse insect groups inhabiting the current plains, comparing their traces with those of the fossil record. Bown (1982) reviewed ichnites and rhizoliths from fluvial deposits of the Oligocene in Egypt, and Laza (1982) described an anthill of the Late Miocene in Argentina. These and other contributions indicated the necessity of assessing different terrestrial environments, organisms that lived in them and the footprints they left. Thus, successive proposals have been added to the continental system of ichnofacies, for instance, the Mermia ichnofacies (Buatois and Mángano 1995a) which corresponds to subaqueous tracks, formed by vertical holes in the shape of Y or U, horizontal holes and others in meniscal shape. The original proposal of the Termitichnus ichnofacies (Smith et al. 1993) was reviewed later by Genise et al. (2000) who suggested to assign such ichnofacies to groups dominated by termite nests in paleosols of closed forest ecosystems under certain humidity and temperature conditions (Melchor et al. 2012). The invertebrate trace fossils are found in numerous continental environments, represented by lithofacies with different degrees of preservation (Hasiotis and Bown 1992), many of them originated in subaqueous environments. However, it is in the paleosols where the fossil traces of vertebrates, invertebrates and especially of insects are more abundant and have greater diversity (Genise 1999). In the modern ichnofacial analysis, all evidence (physical, chemical and biological) must be integrated and used in the interpretation (Frey et al. 1990; Fig. 2.1) The great deal of information resulting from the study of paleosols and their ichnofaunas led to the creation of the Coprinisphaera ichnofacies (Genise et al. 2000), an ichnofacies that gathers all the "Seilacherian" qualifications of recurrence in space and time and at the same time involves great ichnodiversity. This group shows an absolute dominance of chambers as well as systems of holes and chambers. The abundance of coprophagous beetle nests (Coprinisphaera), the most numerous components of this archetypical group, gives the name to this ichnofacies. They inhabited the soils (paleosols) developed in systems of plain deposits or with little relief, in communities (paleocommunities) of dominant Gramineae, but also tree-covered zones. Such ichnofacies consists of a moderate to high diversity of traces that includes various nests of coprophagous beetles, honeycombs of bees— solitary or grouped, wasp nests, anthills, termite mounds, meniscal and smooth tubes (produced by different invertebrates), vertebrate caves, as well as root marks, coprolites, gastrolites and regurgitations. The mentioned list comprises great part of the traces that a group of ichnologists proposed to recognize formally as signs (Genise et al. 2004; Bertling et al. 2006).

Recent studies have brought into debate the need of creating new ichnofacies or ichnosubfacies in the field of continental ichnology, recognizing and limiting specific environments (Melchor et al. 2006; Genise et al. 2008a, b).

More recently, Genise et al. (2010) proposed the creation of Celliforma ichnofacies, identified by the uninterrupted presence of bees and wasps activity in paleosols. Such ichnofacies may indicate biotopes with prevalence of dry climates and poor vegetation cover.

The Skolithos ichnofacies (originally defined in the marine domain) has been tentatively recognized in fluvial deposits of channels and lake deposits of high energy. The ichnofauna is commonly mono-specific, dominated by vertical holes, Y or U shaped (Buatois and Mángano 1998, 2004, 2007; Melchor et al. 2003).

The contemporary continental ichnology can be characterized as one of the most dynamic fields within paleontology and with great potentialities for providing crucial evidence in the reconstruction of life history. The tasks have been focused on two complementary research lines; one of them looks for the characterization of recurrent

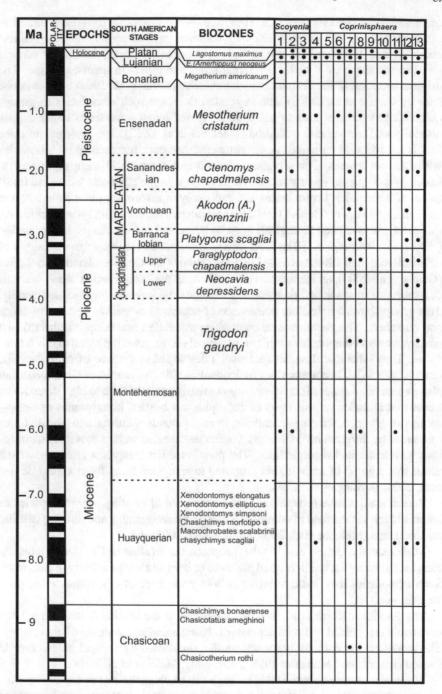

Fig. 2.1 Stratigraphic scheme of the Neogene of the Pampas and the ichnites present in such level (drawing by Marcela Tomeo)

associations with ecological signification and the other one explores evolutionary trends. Thus, the radiation of groups of organisms, the colonization of ecospaces and the appearance of new behaviors are studied (Mángano and Buatois 2001).

In contrast, the proposals of vertebrate ichnofacies, or the incorporation of vertebrate trace fossils for unifying the ichnofacies scheme have been controversial. The principles used for recognizing potential ichnofacies with vertebrate tracks are not the same as those applied for establishing ichnofacies with signs of invertebrate activity. These differences have been recognized and have been suggested for identifying two kinds of ichnofacies: (a) those ichnofacies that reflect interacting behaviors between organisms and substratum (ethoichnofacies) and (b) those that deal with the relationship of tracks and traces with the producer taxonomy (biotaxo-ichnofacies) (Hunt and Lucas 2007). The vertebrate ichnofacies found in fluvial deposits include those of Grallator and Batrachichnus (Hunt and Lucas 2007; Lockley 2007; Lockley et al. 1994).

Although the studies on vertebrate coprolites and the application of ichnotaxonomic names are still in its early stages, coprofacies have been recognized (Hunt et al. 1994), being suggested those of Heteropolacopros and Dicynodontocopros for fluvial sequences (Hunt et al. 1994). The first observations about fossil insect nests in Argentina were carried out by a French researcher, August Bravard (1857a). This author mentioned that in the "toscas (i.e., pedocalcic paleosols) del Río de La Plata" (Pliocene to Early Pleistocene in age), the finding of cells of dipterous chrysalis (which he attributed to the genus Athericea), appearing in a great number surrounding articulated skeletons of fossil mammals, even when the dipterous puparia may have been a fossil body rather than a trace. Ameghino (1880) confirmed such findings in the Luján River basin, Buenos Aires province. His observations were repeated in the coastal outcrops of the cities of Mar del Plata and Miramar, where he mentioned the finding of solitary bee nests such as Ancyloscelis anales Vach in Pliocene–Pleistocene horizons, included in fragments of the "tierras cocidas" ("cooked lands," Ameghino 1908).

In Uruguay, Serafín Rivas (1900) reviewed hymenopteran nests in old deposits, and in Europe, Schutze (1907) described the first fossil bee cell. In turn, the Russian researchers, pioneers in soil and paleosol studies, when mentioning the finding of small caves of fossorial vertebrates, called them "krotovinas" (mole caves, in Russian) (Subachev 1902), name that was generalized until becoming synonym of excavations made by vertebrates.

During the 1930s, several authors gave impulse to these topics in America. In Argentina, Frenguelli (1930) cited fossil bee nests. Frenguelli (1938a, b; 1939b) published the first descriptions of beetle nests which appear profusely in diverse levels of the Tertiary of Patagonia and the Quaternary of the Pampas, relating these nests to "ancient soils." Simultaneously, in Uruguay, Roselli (1939) disclosed fossil nests of beetles and bees of the Early Tertiary in this country, using binomial nomenclature for the trace denomination, as it is used nowadays.

In North America, Brown (1934) created the ichnogenus Celliforma for designing fossil bee cells, publishing both papers about the same topic in 1925 and 1941.

Later, such topics were rarely mentioned and, only recently, the insect trace fossils in paleosols began being studied systematically in order to incorporate them to the theoretical body of ichnology (Genise 1999).

The scientific activity on these fields then continued with Estrada (1941), Sauer (1955), Andreis (1972) and a myriad of works during the following years: Laza (1982), Bown and Kraus (1983), Retallack (1984), Laza (1986a, b), Ritchie (1987), Bown and Laza (1990), Retallack (1990), Hasiotis et al. (1993), Genise and Bown (1994a, b), Dubiel and Hasiotis (1994a, b), Trackray (1994), Hasiotis and Dubiel (1995), Laza (1995), Genise and Hazeldine (1995), Genise and Bown (1996), Hasiotis and Demko (1996), Bown et al. (1997), Genise (1997), Laza (1997), Genise et al. (1998), González et al. (1998), Genise (1999), Duringer et al. (2000), Melchor et al. (2002–2006), Laza (2006a, b), Genise et al. (2007), Verde et al. (2006), Duringer et al. (2007), Bedatou et al. (2008), and Cantil et al. (2013), among many others.

The nesting activity of insects on the soils has an extended colonizing capacity; thus, the sediment removal and the incorporation of organic matter make them great soil makers (Retallack 1990).

Great part of the insect trace fossils in paleosols are nests or part of them, that is, structures excavated and/or built by adults for the progeny development. This activity was called Calichnia by Genise and Bown (1994a) in the ethologic classification of ichnites, being added to the previous ones proposed by Ekdale et al. (1984).

Later, Genise et al. (2007) distinguished a new category for the pupation chambers of insects, which they called Pupichnia. The nests have walls, floor coverings and other quite elaborated devices that offer several diagnostic signs for identifying their builders. The need for maintaining special conditions inside the nests leads the adults to use, often, diverse types of organic matter in the construction, providing them with a marked potential of preservation, since the transformation of such materials favors the concentration of salt and oxides increasing the strength of such nests and their subsequent conservation in the soils (Janet 1898). Thus, these hardened constructions, more compact than the soil surrounding them, have greater possibilities of preservation by diagenesis (Genise and Bown 1994a). The morphological variety of such nests reflects the great ichnodiversity that the fossil record keeps, as well as the various reproductive strategies of the different groups. These characters that in some cases were studied through the reconstruction in 3D by computer (Genise and Hazeldine 1998), computerized tomography (Fu et al. 1994a; Laza et al. 1994; Genise and Cladera 1995) and micromorphological analyses (Cosarinski et al. 2004; Zorn et al. 2010). The study of those strategies comprises the dispersion area of the builder, the choice of the environmental conditions of the nest to be built and the food foraging for the offspring, and its characteristics as well, the inherent climatic factors and the coeval vegetation.

The recognition of these relationships has allowed, for instance, paleogeographic and paleoclimatic inferences about fossil nests of ants and termites (Laza 1982, 1995; Bown and Laza 1990, Genise 1997). In turn, the insect trace fossils in paleosols provided the greatest contribution from the paleoichnology to paleoentomology, thanks to the degree of reliability that the possibility of assigning them to define taxons, such as families and genuses, bring about (Genise 1999). The identification of various and numerous insect groups in paleosols of Pampasia allowed extrapolating many of their trophic activities. Therefore, there are cases related to diverse nests of inquilinism of coleoptera Scarabaeinae, the genus Onthophagus and Canthonini, in natural and mammal caves, as well as remains of Squamata in an anthill of Acromyrmex. Activities of granivorism and mirmecoria produced by ants of genus Pogonomyrmex and Pheidole, cannibalism of ants, genus Forelius, and the cases of mutualism of Attini ants with the fungi they cultivate, have also been reported. Coprophagy activities developed by Scarabaeinae, Aphodiinae and Amitermitinae termites, mycetophagy and necrophagy carried out by Scarabaeinae and dermestids, as well as by the ants Solenopsis and Pheidole, have been observed. The revision of the ichnogenus Coprinisphaera (Laza 2006b) through a very important collection allowed discovering a varied amount of cases of inquilinism and predation of insects in nests of such coleoptera and their subsequent occupation as well (Mikuláš and Genise 2003; Sánchez and Genise 2009). A new revision of ichnogenus Coprinisphaera (Sánchez 2009) added new and important data to its taxonomy, adopted in the present work.

Nowadays, most part of the insect trace fossils may be assigned to three groups: beetles (Coleoptera), termites (Isoptera) and bees, wasps and ants (Himenoptera). Each one of these big groups presents its own peculiarities and requires specific terminology and ichnotaxobases (Buatois et al. 2002). Meeting these needs, Genise (2000) created the ichnofamily Celliformidae, adding then Pallichnidae, Krausichnidae and Coprinisphaeridae (Genise 2004). Recent studies have brought into question the necessity of creating, in the field of continental ichnology, new ichnofacies or ichno-subfamilies, recognizing and limiting specific environments (Melchor et al. 2006; Genise et al. 2008a, b, 2010) (Fig. 2.2).

2.2 The Ichnological Record in the Cenozoic of Pampasia

2.2.1 Vertebrate Footprints (Fig. 2.3a and b)

Most vertebrate footprints stem from continental environments, marginal to the marine environment such as those of the domain of Scoyenia ichnofacies, or in continental waters of the Mermia ichnofacies. These are environments with a higher content of moisture or which are even partly or fully flooded.

As a historical anecdote, it should be mentioned that the first mention of the finding of fossil vertebrate tracks in South America was in 1839 in Colombia (Buffetaut 2000).

Casamiquela (1983) was the first mentioning findings of vertebrate tracks in the Pampasia area. The first finding was in the Río Negro Formation (Late Miocene of northern Patagonia), close to the mouth of the Río Negro (41° 04′ S–62° 47′ W) and the second one was in Late Pleistocene sedimentary rocks, nearby the locality of

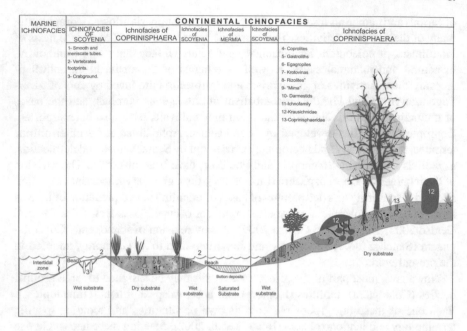

MARINE ICHNOFACIES	CONTINENTAL ICHNOFACIES					
	ICHNOFACIES OF SCOYENIA	Ichnofacies of COPRINISPHAERA	Ichnofacies of SCOYENIA	Ichnofacies of MERMIA	Ichnofacies of SCOYENIA	Ichnofacies of COPRINISPHAERA

1- Smooth and meniscate tubes.
2- Vertebrates footprints.
3- Crabground.

4- Coprolites
5- Gastroliths
6- Egagropiles
7- Krotovinas
8- Rizolites"
9- "Mima"
10- Darmestids
11-Ichnofamily
12-Krausichnidae
13-Coprinisphaeridae

Fig. 2.2 Scheme of continental Ichnofacies in the Neogene of the Pampas (drawing by Marcela Tomeo)

(a)   **(b)**

Fig. 2.3 **a** Human footprint at (Monte Hermoso); **b** Mammal tracks (Pehuén-Có locality)

Monte Hermoso (38° 47´ S–61° 53´ W) corresponding to the edentates Megatheriidae and Scelidotheriinae, respectively.

Aramayo and Manera de Bianco (1987a, b, 1 and 2) discovered the ichnite site of vertebrates and invertebrates of Late Pleistocene age to the east of the Pehuén-Có locality (38° 56´ S–61° 53´ W) dated at 12,000 ± 110 ^{14}C years B.P. This finding spread to a new site, with two sectors, 4 km to the west of the mentioned town (the Playa del Barco site) with an age of 16,440 ± 320 ^{14}C years B.P. These sites gave rise to diverse publications where Xenarthra, rodents, carnivores, Liptorerna, Prosboscidea, Perissodactyla and Artiodactyla mammals were identified, and among the birds, Phoenicopteriformes, Tinamiformes, Anseriformes and Rheiformes (Aramayo and Manera 1996, 1998, 2000), Aramayo et al. (2003), Manera and Aramayo (2004). Later, Manera et al. (2005) distinguished traces of animal skin in the flange of tracks attributed to Megatherium. In 2005, isolated human tracks were found in the same site, which were dated in 12,000 radiocarbon years B.P. (Aramayo and Manera 2009; Manera et al. 2010).

Quintana et al. (1998) mentioned the presence of mammal ichnites imprinted on sediments of Late Pleistocene age, in the Burucuyá cave, western slope of the Sierra de la Vigilancia (Buenos Aires province). This finding spread then to Gruta del Oro, both sites located in the Sierras de Tandil area.

Aramayo (1999) discovered tracks that were assigned to Promacrauchenia and small rodent crotovinas in the "Rionegrense" sediments of the Atlantic Ocean coast, 30 km westward of the Río Negro mouth. Sometime afterward, Aramayo et al. (2004a, b) found tracks of xenarthrans, ungulates, carnivore marsupials and phororhacoids birds in coetaneous deposits (between the El Cóndor and La Lobería beach resorts; province of Río Negro, 41° 3′ S –62° 53′ W). The sediments corresponding to the Late Miocene-Early Pleistocene epochs were typified as deposits belonging to interdune temporal lagoons, with desiccation crevasses and ripple marks.

Tassara et al. (2005) informed the finding of vertebrate ichnites in the Santa Clara Formation (Late Pleistocene), in the coastal sector of Barrio Parque Camet Norte, in Mar Chiquita county and in the mouth of Arroyo Seco (37° 47′ S–57° 27′ W), both sites in Buenos Aires province.

They recognized four sectors, carriers of the ichnogenuses already mentioned in the Pehuén-Có site.

In sediments of the Early Holocene, 6 km west of the Monte Hermoso locality (southern Buenos Aires province), several hundreds of human footprints, some others of birds and an artiodactyl footpath, were found. The footprints are distributed sporadically around 800 m length at the present coastline (Politis 1993).

Aramayo et al. (2007) pointed out that tracks belonging to tardigrades xenarthrans—adult and young—equids and artiodactyls, which were assigned to the Late Pleistocene, were found in the Monte Hermoso locality (the Camping Americano site, Buenos Aires province).

2.2.2 Coprolites (Fig. 2.4)

Coprolites are fossilized feces (of vertebrates and invertebrates) which underwent burial processes and subsequent mineralization for having been deposited in calm flooded zones or in drier zones, where they received fast burial. They can be identified by their morphology, extrusion marks, sutures, gas bubbles and inclusions

Fig. 2.4 Coprolites

such as food waste (either animals and/or plants), being also pollen and phytolith carriers. They can sometimes provide information about the taxonomic identity of the producing organism, due to their shape.

In paleosols of the Cerro Azul Formation (Huayquerian stage, Late Miocene) from the Caleufú locality, province of La Pampa, Argentina (35° 41´ S–64° 40´ W), coprolites of assumed predatory vertebrates were collected that contain microvertebrate remnants. These remains are modified by chewing and digestion and are quite difficult to identify. The materials agglutinated in cemented sand by calcite are of cylindrical shape, 14–16 mm in diameter and 8–21 mm long (Montalvo 2004).

Aceñolaza and Aceñolaza (2004) indicated that numerous vertebrate coprolites had been found in Ituzaingó Formation, in outcrops on the Paraná River banks. Noriega and Areta (2005) pointed out the presence of herbivore coprolites together with remains of insects, plants, ostracods and osseous relics of the huge condor bird Sarcoramphus papa, in the section of Camet Norte locality, level B (Lujanian stage, Buenos Aires province).

Martínez and Ubilla (2004) mentioned the finding of coprolites with tooth and rodent bone inclusions in paleosols of the Sopas Formation (Lujanian stage) in northern Uruguay, venturing that such remains were the result of carnivore activities.

In the Ensenadan stage, sediments from the town of Miramar, Buenos Aires province, a coprolite was found with inclusion of Lagostomus bones, associated with glyptodont remains; this was the reason why the coprolite was attributed to a carnivore mammal.

The presence of feces, attributed to a canine mammal in the Lujanian stage sediments (the Guerrero Member, Late Pleistocene), has been mentioned for the locality of General Guido, Buenos Aires province (Chimento and Rey 2008).

2.2.3 Gastroliths, Enteroliths or Bezoars (Fig. 2.5)

Rounded stones, commonly selected and ingested by reptiles and birds, which help such vertebrates in the inner food milling, are called gastroliths. Their finding was not mentioned in the different levels of the Cenozoic units in Pampasia. The enteroliths of bezoars are formed inside the digestive tract of mammals. These structures can be organic or inorganic; those of organic origin may be formed by fur ingestion (trichobezoars) or plant materials (phytobezoars). Only one finding has been recorded for the region: It corresponds to an accretionary body found by Florentino Ameghino in the Late Pleistocene of the Olivera locality, Luján County, Buenos Aires province. The box where these materials were stored was found many years later in the La Plata Museum; it had also osseous remains of Scelidotherium. It is an ovoid body, partly embedded in "tosca" (Ca carbonate), of 62 × 89 mm in size. Its surface is rough, bright and ocher brown; the analysis of the cuts by diphractograms allowed to identify it as a fossil phosphatic stone (Teruggi et al. 1972).

2.2.4 Egagropiles (Fig. 2.6)

Regurgitations or pellets are waste resulting from the predating activity of several birds, such as Strigidae, Tytonidae, Laridae and Ardeidae, which usually inhabit big trees and places such as caves, caverns or natural cavities. Birds swallow their whole preys—small mammals, fishes, mollusk, birds and insects—whose waste is eliminated in the shape of bolls (pellets). Pellets are characterized for keeping the bones hardly fractured and the cranial bones disarticulated, contributing to the small vertebrate record and thus supplying the fauna sampling in a certain area. The paleontological record mentions them rather frequently calling them "microbonebeds" (Terry 2004). Frenguelli (1928) was the first author to mention them for the coastal cliffs of Buenos Aires Province. Tonni and Fidalgo (1982) mentioned their presence in the outcrops of Punta Hermengo, Miramar County, Buenos Aires province.

Tonni et al. (1993, 1998) informed the finding of two regurgitation accumulations of Strigiforme birds that provided several complete pellets with micromammal and bird remains; the first forming part of the filler of a cave of great edentates located to the SW of the Punta Hermengo locality, General Alvarado County, Buenos Aires province (38° 16′ S–57° 50′ W). The second finding was in the marine cliffs to the N of the city of Mar del Plata (37° 56′ S–57° 32′ W), General Pueyrredón County, Buenos Aires province. Both stratigraphic levels were correlated, and they correspond to the Late Ensenadan stage (Early to Middle Pleistocene).

Fig. 2.5 Gastrolith, enterolith or bezoar

Mazzanti and Quintana (2001) described pellet layers in the filler of the archeological site Cueva Tixi, in Sierra de la Vigilancia, eastern Sierras de Tandil, General
Pueyrredón County. Levels D and E have accumulations of pellets, and they were
dated in 3255 + - 75 ^{14}C years BP and 10,375 + - 90 ^{14}C years BP, of Holocene
and Late Pleistocene age, respectively. Laza (1998, 2001) stated that the pellet accumulations of Cueva Tixi, as well as bird excretions producing them, and probable
deposits of bat excrement created a substratum quite favorable for the establishment
of beetle populations of the genus Onthophagus, whose nests are associated with
such accumulations.

2.2.5 Krotovinas (Fig. 2.7a, b, and c)

The first mention about the presence of krotovinas in the Pampasia sediments corresponds to Florentino Ameghino (1880) when pointing out vertebrate caves attributed
to vizcacha (a large South American rodent), foxes, mice and tuco-tuco (a smaller
rodent) in the Luján River basin, Buenos Aires Province. In 1908, while exploring

Fig. 2.6 Egagropiles

the Atlantic Ocean coast between the Miramar and Mar del Plata counties, Ameghino observed large and small filled caves at different levels of the Pliocene–Pleistocene sediments, which he assigned to mammals.

Frenguelli pointed out the existence of such caves in his 1921 and 1928 papers arguing that the largest ones prevail in deposits of the "Pre-Belgranense" sediments, belonging to the Marplatan–Ensenadan stages, and attributing them to glyptodonts. This argument was unshared by Kraglievich (1934), who believed that these xenarthrans were unable to dig due to their shell stiffness and the lack of frontal legs suitable for such a task. Frenguelli also mentioned a cave filled with volcanic ashes with remains of Scelidotherium in the locality of Centinela del Mar, Buenos Aires Province (38° 25′ S–58° 09′ W). Rusconi (1937e) attributed these blind caves to the activity of large dasypodidae. In 1967, he wrote that "winding strips of about one meter wide"…"are formed by clayish materials and arranged in small layers up to one mm thick. These materials differ from the adjacent loessoid material" at the surface of the Ensenadan stage sediments of the river banks north of Buenos Aires City (the town of Olivos), in times of lower water level of the Río de La Plata. He also verified that, in some places, these sediments outstand from the surface and, in others, they appear at a lower level due to different diagenetic qualities, illustrating schematically these findings and others in the sediments of the Bonaerian stage in the city of Buenos Aires.

Kraglievich (1952) revealed caves up to 1 m in diameter within the level II of the Vorohué Formation (Marplatan stage). Imbellone and Teruggi (1988) and Imbellone et al. (1990) mentioned that a krotovina was found, formed by sediments of the

(a)   (b)

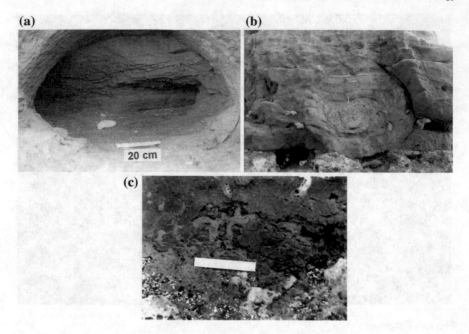

(c)

Fig. 2.7 Crotovines (**a**, **b**, and **c**)

Bonaerian stage, in a quarry from Gorina, a locality near the city of La Plata (Buenos Aires province, 34° 54´35" S–58° 00´15" W).

Genise (1989) published the first paleontological observations about caves with Actenomys and associated fauna belonging to the Chapadmalal Formation. He studied 250 caves in Barranca de Los Lobos, Las Palomas beach, Los Lobos, Chapadmalal and Barranca Parodi sites, Buenos Aires Province. The caves mostly had remains of Actenomys and other rodents, notoungulata and small marsupials, in some cases articulated. The caves with Actenomys, of 13 cm in diameter, penetrate in the ground with a slope from 30° to almost vertical and a length up to 5 m. They have lateral ramifications; widening for the displacement in both directions and the possibility of more than one chamber per cave (see associated Scarabaeinae nests). Scognamillo et al. (1998) carried out morphometric studies about the same cave systems with Actenomys in the Barranca de Los Lobos zone, confirming the aforementioned data.

Nabel et al. (1990) carried out geological studies in the San Pedro and Baradero localities.

Their profiles mentioned "one krotovina of 1.50 m in diameter" in the upper part of the basal level (Ensenadan stage) and "krotovinas and bioturbation" in the upper level (Bonaerian stage). Zavala and Navarro (1993) pointed out the presence of krotovinas from 18 cm to 1 m in diameter and up to 4 m long for sediments of the Montehermosan stage, some of them with ramifications, which in the end present significant widening and inclinations of 7°.

Quintana (1992) described a cave found during excavations in the city of Mar del Plata, together with other three findings of similar characteristics. Excavated in sediments of Pleistocene age (the Miramar Formation, Ensenadan stage), the cave consists of two interconnected galleries of 23 and 3 m, with a sealed end, of semicircular section and a plain base of 0.93 m wide and 0.76 m high. Later, Genise and Farina (2012) redescribed such cave together with another one mentioned by Dondas et al. (2009). In such work, the idea, profusely documented, that those caves express foraging activity with their extended and horizontal development is added to the previous observations, since in both examples, the excavations cut lands bearing ant nests, the latter appearing being excavated. The caves show deep unguiform marks performed by the builders, probably a large armadillo (Eutatus, Propraopus, Pampatherium). Their filler contains a big deal of arthropod remains, especially insects, such as ants (Neivamyrmex sp., Acromyrmex sp., Pheidole sp. and Solenopsis sp.), Rhyparochromidae, Tenebrionidac, Staphylinidae, Histeridae, Scarabaeidae, Aphodiidae, Carabidae, Cydnidae, Enicocephalidae and Termitidae. The present author observed caves assignable to rodents with constant diameters of 20 mm presenting vertical and horizontal developments communicated by very pronounced sinuosity in gullies that extend to the south from the site of Punta Hermengo, at levels corresponding to the Ensenadan stage.

Zárate et al. (1998) cited large caves in the Mar del Plata region, Buenos Aires Province, pointing out the presence of footprints of their builders on walls and roofs. The diameter varies from 0.75 to 2 m in the 42 measured caves, some partially filled. Many of the structures go beyond these dimensions, with ramified tunnels excavated up to 5 or 6 m deep, reaching an extension of 20 m. The caves extend throughout all the Cenozoic sequence, and their builders could be Xenarthra, large dasypus and pampaterids as well as Mylodontidae (Scelidotherium and Glossotherium). The studied sites correspond to the localities of Santa Isabel Beach, Baliza Caniú—Serena Beach, Colonia Chapadmalal, Constitución—Camet and Cantera Vialidad.

Vizcaino et al. (2001) added new observations to the mentioned discoveries thanks to anatomic and biomechanical studies of Pleistocene mylodontidae that indicated that their members were well-provided for digging. Iriondo and Krohling (1996), when describing the Carcarañá Formation (Lujanian stage) from Santa Fe province, Argentina, mentioned krotovinas of 0.50 to 0.75 m in diameter in two levels, one of them containing remains of carbonized plants.

De los Reyes et al. (2006) have stated that they found a peculiar taphocenosis associated with a paleocave formed by remains of marsupial Thylophorops together with others of dasypus, ctenomydae and caviidae, and also by coprolites of a possible carnivorous, at levels IX or X of the Chapadmalal Formation, 200 m to the south of the Arroyo Las Brusquitas mouth, General Alvarado County. Such coprolites have ground bony remains assignable to individuals of mentioned fossorial taxa. The authors point out the possible reuse of eutatin caves by marsupials. Dondas et al. (2009) described caves in the area of Mar del Plata, coming from the Miramar Formation, identifying three kinds of galleries: (1) the largest attributed to Glossotherium; (2) somewhat smaller ones assigned to Scelidotherium and (3) assigned to Pampatherium. They recognized the digit marks on the excavation walls for differentiating between mylodontides (two digits) and dasypus (three digits). In one of such caves, they found mounds similar to those attributed to Attini ants by Laza (1982).

Verde and Ubilla (2002) mentioned the presence of krotovinas in the Sopas Formation (Late Pleistocene of Uruguay).

In Brazil, findings of krotovinas were mentioned by Paglarelli Bergqvist and Maciel (1993, 1–2), in the coastal plain of the state of Rio Grande Do Sul (Tapes and Vila Cristal counties). They correspond to large caves in the Lateritas Serra de Tapes Formation and others in the Graxaim Formation, both of Pleistocene age. Three of them have 1 m in diameter, and the other one has an elliptical contour of 0.55 by 0.72 m. They were attributed to cingulates xenarthrans such as Pampatherium, Holmesia and Propraopus. In addition, mammal krotovinas were found in the system Barras Litorales I, also attributed to large dasypus (Buchman et al. 2009). Elorriaga and Visconti (2002) pointed out the presence of krotovinas from the Cerro Azul Formation (Huayquerian Stage, Late Miocene) in road cuts of the National Route 154, southeastern La Pampa Province, Argentina. They are sub-circular from 1.0 to 1.20 m high and 1.0 to 1.60 m wide; the largest one is 3.0 m wide and 0.80 to 2.10 m high (one of them is 9.0 m long). The filler, passive, has plant remains. They were attributed to Scelidotherium or Glossotherium, sensu Vizcaíno et al. (2001). Zárate et al. (2007), while describing the stratigraphy of the Chasicó Formation, mentioned the presence of excavations of 30 cm in diameter without further description.

The present author collected several tiny Scarabaeinae nests (a finding that will be further discussed when dealing with the Coprinisphaeridae Ichnofamily) in the Chapadmalal Formation, in the gully sector called Bajada Martínez de Hoz (Mar del Plata; 38° 08´ S–57° 36´ W), having paleosol levels with numerous caves with Actenomys, one of them forming part of the filler.

2.2.6 Rhizolites (Fig. 2.8)

The root traces are evidence that the rock housing them was exposed to atmospheric phenomena and colonized by plants, thus characterizing a soil. They are called rhizolites and are defined as organic-sedimentary structures; they have been known since the beginning of last century (Todd 1903). The presence of such rhizolites was considered by Andreis (1981) and Retallack (1988) as one of the main attributes of a paleosol. Klappa (1980) has mentioned that the field and petrographic observations indicate that the roots of superior plants produced numerous shapes characteristic of calcretes, which he ordered in five basic types: molds, casts, tubules, root casts s.s. and petrifactions. Kraus and Hasiotis (2006) recommended, in the study framework of the climatic conditions of the past and the environment reconstruction, careful observations and geochemical analyses of rhizolites that help in the interpretation of ancient drainage conditions.

In floodable plains with meandering fluvial systems, developed in arid to sub-humid climates such as those developed in Pampasia, intercalations of continental

Fig. 2.8 Rhizolites

limestones carrying rhizolites, sometimes associated to hymenopteran nests (Celliforma), are usually frequent denoting the humidity degree of the plains (Plá et al. 2007).

Frenguelli (1926) related the calcium carbonate casts deposited on roots of living plants with the structures that existed in the Pampean loess. Due to this, he classified them as root casts. Despite the previous records, not much has been studied about the shape and development of such fossil structures and the reference about the presence of rhizolites in the description of numerous paleosols in Pampasia are erratic and infrequent.

2.2.7 Crabs and Their Signs of Activity (Fig. 2.9a, b)

The continental territory periodically flooded by the sea includes a particular habitat where organisms associated with periods of aquatic and sub-aerial life prevail. This habitat was typified as a Scoyenia ichnofacies settlement. One of the most conspicuous groups that inhabited this territory is formed by crabs. Such group shows a wide range of adaptations to the habitat, occupying continental and sea environments of marine and continental waters. The locomotion, excavation and feeding activities of benthic crabs are most appropriate for providing ichnological records, and their caves are abundant in numerous sedimentary environments, from terrigenous sediments to shallow marine environments with abundant carbonate precipitation.

The digging crabs show two behavioral patterns in their task: backward and sideways (Frey et al. 1984b). The burrowed sediments usually form spherical pellets, which once removed from the hole are then deposited outside. Some species use pelletized sediments in the wall construction, extending the duct vertically and horizontally, granting a mamelon texture to the outer surface by the adherence of such pellets.

Fig. 2.9 Signs of activity of
crabs (a, b)

The ichnogenuses recognized in the region under study correspond to two forms assignable to constructions of individuals of Callianassa genus, whose characteristics were described by Frey et al. (1984b).

Psilonichnus: These are living spaces formed by predominantly vertical J-, Y- or U-shaped holes, of variable diameter, with lateral branches that are present forming singular or forked cul-de-sac, tending to verticality in a chimney shape. They are related to supralittoral forms; estuary, beach and sand dune dwellers (Nesbit and Campbell 2005).

Ophiomorpha: These are typified as branched tubes with a tridimensional horizontal, slanted or vertical lattice. The external part of the tube is characterized by a bulbous texture due to the gallery reinforcement with pellets. It is interpreted as a feeding and housing construction of a decapod crustacean (Callianasidae). The genus revision carried out by Bromley and Ekdale (1998) referred it to shallow marine environments (barriers from frontal deltas with silt to fine sand sedimentation).

The sea transgressions of the Pleistocene epoch left diverse testimonies of their invasion of the continental margin along the Atlantic Ocean coast of the region under study and linked to these deposits, footprints of the settlement of crab populations are recognized.

As it follows below, there is a record of these deposits:

1. Deposits corresponding to the "Belgranense" marine transgression interposed between the continental sediments of the Ensenadan and Bonaerian stages (Early to Middle Pleistocene) (Cione et al. 2002). They emerge in Santa Clara del Mar, north of the city of Mar del Plata, where ichnites assignable to the ichnogenus Psilonichnus coming from the settlement of crab populations were found.

2. Mouzo et al. (1985) pointed out that a sand layer with tubes cemented by calcium carbonate (Ophiomorpha) from 4 to 10 cm in diameter and between 10 and 50 cm long appeared along 8 km on the beach of Pehuén-Có; they have rough external walls and smooth inner walls, which they attributed to callianassidae. Frenguelli (1928) stated, while describing these outcrops: "The origin of these

pellets is not very clear. They consist of a very sandy and light limestone, and of light grey sandstone forming long irregular cylinders, ramified, of 5 or 6 cm in diameter, solid and excavated tube shaped. The filled cavities probably correspond to burrows belonging to small rodents" … "the petrographic characters of the calcareous sandstone filling them, … often have moulds of small marine bivalves". In the same layer, he found fragmented remains of Paraceros, Equus, Mastodon, Toxodon and Scelidotherium. The same layers were called "Secuencia San José" by Zavala and Quattrocchio (2001), and in its Lower Section (Early Pleistocene), they described Ophiomorpha nodosa bioturbations in life position on thick layers, as well as associated marine fauna. Such environment may correspond to that of a shallow sea in transgression, with non-canalized, high-density irruptions of fluvial currents. Similar deposits were recorded in Uruguay, in the Colonia and Rocha counties. During the Holocene, the marine transgression named as "Querandinense" developed several crab habitats. Rossi et al. (2001) indicated that the stratigraphic sequence in Mar Chiquita (Santa Clara Formation, Late Pleistocene) is intensely bioturbated. Tracks of tubuliform excavations with or without meniscus filled with pellets or sediments prevail. They recognized the ichnogenuses Taenidium, Skolithos and Edaphichnium and also mentioned nests of social insects, systems of arthropod galleries and others not distinguishable, assigning the sequence to the Scoyenia ichnofacies. It is worth pointing out that in these sectors, there were printed ichnites corresponding to two ichnofacies: (a) the soil development during the Lujanian stage, which includes ichnites corresponding to Coprinisphaera ichnofacies; (b) outcrops then covered by the sea during the "Querandiense" marine invasion, when the Scoyenia ichnofacies, carrier, among others, of crab habitat construction tracks, developed. Osterrieth et al. (2004) described a similar situation when presenting the finding of a fossil crab habitat about 8 km to the SW of Mar Chiquita, in Los Cueros Stream, which developed on paleosols dated in 9516 ± 512 ^{14}C years B.P. until around 3950 B.P. The scenery may represent a lagoon estuarial environment. The design and dimension of paleocaves, as well as the found remains point out the presence of Chasmagnatus granulata (Aramayo et al. 2005). The latter authors, while studying the coastal evolution between Monte Hermoso and Pehuén-Có localities, distinguished the succession of three environments for the Holocene:

a. Continental, formed by lenticular silty-sandy bodies.
b. Beaches and coastal lagoons, with lagoon-like deposits and caves with remains of the Chasmagnatus granulata crab that does not live in the area at present.
c. Marine, represented by a platform of sandy-clayey sediments.

2.2.8   Activity of Dermestids

Remains of fossil vertebrates with marks of activity of such coleoptera exist in the La Plata Museum collections. Martin and West (1995) placed these signs of activity in the Cubichnia classification (sensu Seilacher, 1953), whereas Roberts et al. (2007) created the taxon Cubiculum ornatus. Ethological observations showed that insects feed on the vertebrate carcasses exposed to the open air, on dry tissues; they develop their oviposition digging small ovoid cells on the bones where the larvae pupation takes place. Most publications on the subject are based upon the findings related to dinosaur osseous remains affected by the activity of the mentioned coleopteran (Roger 1992; Paik 2000; Britt et al. 2008).

Voglino (1999) mentioned perforations of 2 to 8 mm in diameter that he assigned to dermestids for Paraná gullies in his unit S7 (Bonaerian stage) on Glyptodon, Sclerocalyptus and Morenalaphus bones.

Other insects that predate on osseous remains and dry tissues are the termites (Darry 1911). Recent taphonomic studies in the site Paso Otero, in the Río Quequén Grande, Buenos Aires Province, within Lujanian stage deposits (Late Pleistocene) provided the collections with osseous remains with marks assigned to Isoptera (Pomi and Tonni 2010).

2.2.9   Smooth and Meniscated Tubes (Fig. 2.10a, b)

The anelid activity is recorded very often on soils and sedimentary deposits where they live, and ichnology records these activity signs with the name of Taenidium from which several species are recognized. Taenidium barretti (Bradshaw) was found, associated with levels which recorded numerous and varied vertebrate footprints in the Pehuén-Có locality, Coronel de Marina Rosales County, Buenos Aires Province, in the Guerrero Member of the Luján Formation. The ichnite description corresponds to a cylindrical horizontal winding trace, in part rectilinear, with intersections, without wall and with curved meniscuses, 4 mm thick and 3 mm apart, one from another, lacking fecal components. The trace width ranges between 9 and 12 mm (Aramayo et al. 2005). The possible builders are worms. T. barretti was recorded in alluvial and lacustrine environments, and in flooding plains and fluvial channels (Keighley and Pickerill 1994), a fact verified also by Aramayo and Manera de Bianco (1987a, b). These finding gives these layers the characteristic of being carriers of ichnoforms, corresponding to the Scoyenia ichnofacies in their transit to continental terrigenous environments.

The sites with paleosols of the Cerro Azul Formation (Huayquerian Stage) where Taenidium was found are: (a) Fatraló, Buenos Aires Province, close to the border with the La Pampa province; (b) gullies to the SE of the Utracán shallow lake, in the county of the same name, in La Pampa Province. Zavala and Navarro (1993) pointed out meniscal tubes for the Monte Hermoso Formation (Montehermosan stage) that they assigned to Muensteria (sic) genus. Deschamps et al. (1998) described the biostratigraphy of two sites close to the city of Bahía Blanca, named as "Grumbein" and "Relleno Sanitario Quarry"; in the latter, sediments corresponding to the Chapadmalalan stage, Taenidium associated with roots and vertical tubes were found.

(a)   (b)

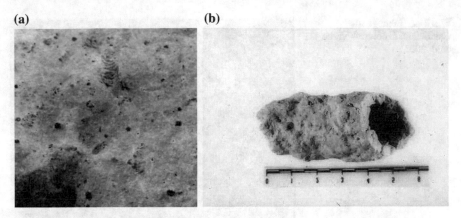

Fig. 2.10   Smoot and meniscated tubes (a, b)

In sediments of similar age (Chapadmalalan stage), specimens of Taenidium were collected in the marine cliffs of General Pueyrredón County, in the Antenas Militares site (Esteban Soibelzon, personal communication).

Voglino (1999) mentioned, for the Río Paraná gullies, meniscal tubes similar to Taenidium in stratigraphic units corresponding to the Bonaerian stage.

2.2.10   *Castrichnus Incolumis Verde et al. (2006) (Fig. 2.11)*

In Uruguay, Tacuarembó, Salto and Artigas counties, sediments of the Sopas Formation (Lujanian stages) including paleosols carriers of diverse Ichnofossils emerged. Among these, Castrichnus incolumis and Taenidium serpentinum appeared associated; this fact was interpreted as a compound fossil trace (Verde et al. 2006). C. incolumis was described as chambers of worm stowage, whereas T. serpentinum could correspond to displacement tunnels belonging to those worms. The chambers range from spherical to slightly ovoid, of 30.3 × 28 mm, with walls formed by interwoven pellet layers and covered inside with concentrically smooth pellets. The chamber filler is formed by pellets of 5.4 mm, arranged in cords attached to the wall.

T. serpentinum appear from winding to straight, with diameters between 0.40 and 0.75 mm; the meniscuses are arched and the distance between them is less than the duct diameter. The external molds show the rings corresponding to the meniscuses. They have well-marked borders, without ramifications. The filler is similar to a rock bearer.

The association of C. incolumis with T. serpentinum indicates the worms as potential producers of meniscal holes in paleosols. Similar structures were found by Elisa Beilinson (2010) in the San Andrés Alloformation (Pliocene-Pleistocene) of Buenos Aires Province. They appear associated with calcium protosols. In such geological

Fig. 2.11 Ichnogenus
Castrichnus incolumis Verde
et al. 2006

sequence, meniscal holes also appeared (Taenidium) with a degree of bioturbation of 4-5 in protosols and Beaconites from 6 to 10 mm in diameter.

2.2.11 *Edaphichnium Bown and Kraus 1983 (Fig. 2.12)*

These structures are cylindrical tubes circularly cut, filled with rounded fecal pellets that give the surface a rough texture. They are horizontal, without ramifications. The finding took place in a quarry in the site of Arturo Seguí, near the city of La Plata, in the Upper Luján Formation, formed by eolian sediments that crown the higher areas in the region. The specimen, arranged horizontally, has a slightly winding extension, and along this extension, it appears to be composed of pellets in some sectors, giving a rough texture. The specimen is 500 mm long and 27 mm in diameter. The description coincides with that of E. lumbricatum of Bown and Kraus (1983), who attributed such structures to oligochaetes. Fragments of Edaphichnium were also found in outcrops of the Buenos Aires Formation, Bonaerian Stage, in the Río Arrecifes, San Antonio de Areco County (34° 22′ S—58° 35′ W), Buenos Aires Province.

2.2.12 *Mima-Type Mounds*

This term corresponds to mounds of two meters high and some ten meters in diameter. Its formation results from—according to several published works—the activity of digger animals, and they were recorded in densities up to 100 mound per hectare.

Fig. 2.12   Ichnogenus Edaphichnium Bown and Kraus 1983

Although they only modify the land surface, its relative topographic importance can be kept during long periods and attest environmental changes. Its formation could be due to the sediment displacement—as a result of the excavation—toward the periphery of the activity center, usually placed on higher sites. This phenomenon, observed in several places in North America and Africa, was associated with different families of digger rodents. In Argentina, these burial mounds were detected in the SE of San Luis Province, south of Córdoba Province and north of La Pampa Province (Roig and Cox 1985-86; Cox and Roig 1986). They are up to 3.5 m high and 25 m in diameter. They appear occupied by the digger rodent Ctenomys; the grass Cynodon stabilized the sandy sediments, accelerating the formation of mounds in a process of few years (Roig et al. 1988). In the Buenos Aires Province, Mima-type mounds were observed at 35 km west of the city of Necochea, on route 228 (Cox et al. 1992). They were up to 1.5 m high and 20 m in diameter, without the presence of digger rodents. In an area of 2 ha, 42 mounds were observed, finding caves and other excavations of Chaetopractus and Dasypus (dasypodids xenarthrans) and Solenopsis richteri anthills in 21 of them. Two current species of Ctenomys live nearby, in sand dunes of the Atlantic Ocean coast, dismissing the fact that these rodents lived in those clay soils, difficult to excavate. The mentioned authors (Cox et al. 1992) suggested that the origin of burial mounds resulted from Solenopsis ant activity through numerous generations. The vegetation on the mounds appeared richer than in the surrounding area; this difference may have been caused by myrmecochory. At the same time, the mounds could have served as a mammal habitat settlement.

2.3   Insect Nests in Paleosols

The knowledge about the architecture of insect nests comes from entomology, and some of its terms are also used in ichnology. In this field, even the most complex insect trace fossils can be divided morphologically in two components: tunnels and chambers. Those primary differences in the behavior, as well as many other specific differences, gave rise to the great morphological diversity of insect nidifications and pupation structures in soils and paleosols, thus offering invaluable taxobases for the classification of ichnogenuses and ichnofamilies. Bromley (1990) listed and analyzed the most common characters used as basis for ichnotaxonomy in four ichnotaxobases: (a) general shape, (b) wall structure type, (c) type of ramifications and (d) filler nature. Below, it will be only mentioned and described the current shapes in the territory under study.

2.3.1 Ichnofamily Celliformidae Genise 2000

These are materials composed of a group of morphologically related ichnogenuses, formed by cells constructed by bees for breeding their progeny and whose type ichnogenus is Celliforma Brown 1934. Celliforma, the simplest trace of the group, consists of chambers or its inner molds of different shape (sub-cylindrical, in the shape of drop, bottle, vase, glass and barrel); they have a rounded end and the other truncated or blocked by a welded seal in spiral shape as part of its structure. The ichnofamily includes shapes with antechambers and discrete walls, as well as structures formed by several cells.

Record in paleosols of Pampasia.

Miocene. Cerro Azul Formation, "Huayqueriense" stage. Arroyo El Venado between Guaminí and Carhué counties, Buenos Aires Province.

Miocene. Monte Hermoso Formation, Montehermosan stage. The longest section is located at the coastal cliffs, coronel de Marina Leonardo Rosales County, Buenos Aires Province.

Holocene. "Platense" (Platan stage). Paraná gullies, Baradero County, Buenos Aires Province. Described by Voglino (1999) as: "A cylinder of 20 mm long by 8 mm in diameter, sealed in an end by a concave-convex plug".

2.3.2 Ichnofamily Krausichnidae Genise 2004

This family is formed by structures ranging from complex to very complex that in general are interpreted as assignable to social insects such as ants and termite nests and whose type ichnogenus is Krausichnus (Genise and Bowm 1994b).

A group of ichnogenuses forms part of this ichnofamily. These are formed by chambers associated by tunnel systems of different diameters in many cases. The tunnels appear without scratches and/or intersected by furrows. The chambers lack irradiation tunnels from their top part and are linked by a duct system that interconnect them with other chambers. These chambers can be empty, passively or actively filled and/or can have secondary system of ducts in different diameters and small chambers on or in the wall (Genise 2004a).

2.3.3 Ichnogenus and Ichnospecies Found in Pampasia

2.3.3.1 Anthills

Ichnogenus Attaichnus Kuenzelii Laza 1982 (Fig. 2.13a, b and c)

The identified remains come from the Miocene, Cerro Azul Formation, Huayque-rian stage, Salinas Grandes de Hidalgo, Atreuco County, La Pampa Province. The structure, complex and huge, forms a burial mound of 7 m in diameter by 3 m high, consisting of countless globular chambers with shapes and sizes comparable to those of anthills of the current species of Atta genus. The chambers of fungus-growing ants are connected to one another and to outside through two systems of ducts: longer and shorter. The longest ones have access to chambers, generally from their bottom part and form a folded flange within the chamber. The longest intercommunication ducts or galleries have sub-oval diameters between 17 and 20 mm. The shortest ducts interconnect the longest chambers and ducts with diameters ranging between 5 and 10 mm. The dimension of the longest and shortest ducts does not vary. The chambers are globular with somewhat irregular inner and outer surfaces. Both larger diameters range from 140 to 170 mm. The walls, whose thickness is notably irregular (between 15 and 50 mm), define a considerably ample globular cavity. The calcium carbonate may be deposited on the fungus-growing ants inside the chambers. One of the chambers presents in its inner part a tubular cavity or turret with edges somewhat folded in lip-shape. Except for the largest dimension in the fossil specimen—the description coincides totally with those by Gallardo (1916) and Bruch (1917) when describing the Trachymyrmex pruinosus Emery nest belonging to the Attini tribe. It is possible that this character is common to the most evolved tribe members in the nest construction. Possible builders of these structures are the aforementioned characteristics allow the classification of the remains as belonging to ant nests of the genus Atta Fabricius.

   After more than forty years, the sediments originally carriers of A. kuenzelii, lost by erosion, recent field and laboratory works cast doubts on the original ascription, attributing anthill of Acromyrmex or Trachymyrmex (Genise et al. 2013) species to Attini. Besides, the idea of sympatry of the mentioned genuses is discarded, not considering that the current populations of the Chaqueña Sub-region, (sensu Morrone 2000–2001) register more abundant nests of Acromyrmex and Trachymyrmex than those of Atta. The work in question does not take into account the paleoecological information supplied by the vertebrate fossils, especially those corresponding to rodents found in different sites of the Cerro Azul Formation in the La Pampa Province.

   Record in paleosols of Pampasia: Other specimens were found in the car track placed 7 km to the south of the city of Macachín, La Pampa Province, in a horizon of the same age, carrier of numerous "slag and baked clays."

(a)   (b)

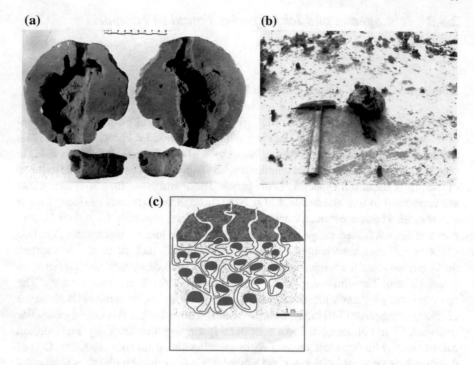

(c)

Fig. 2.13 Ichnogenus Attaichnus kuenzelii Laza 1982 (**a, b**) **c**, scheme

Ichnogenus Aff. Acromyrmex (Acromyrmex) Close to A. Ambiguus Emery
(Fig. 2.14a, b)

The materials come from the site between "Club Náutico" and "Los Tamarindos" beaches, General Alvarado County, Buenos Aires Province. Pliocene, Chapadmalal Formation, Chapalmalalan stage. In that place, it ends in a carbonated horizon (part of a paleosol). The construction has two chambers in an oval shape, totally replaced by "tosca" (i.e., caliche). The upper chamber is larger, 120 mm high and smaller diameter; the lower one is 70 mm high by 60 mm in diameter. Both chambers are 40 mm from each other. The largest chamber shows the inlet and outlet ducting in the shape of inner bubbles (typical of the ichnogenus), upper and lower, while the duct of intermediate connection, replaced by calcium carbonate, is 12 mm in diameter. The hosting rock, when cut, showed sections of ducts filled with dark clay. These ducts, whose sections vary between 5.5 and 9.5 mm, become numerous toward the upper part. Long and short ducts converge at the top of the structure, forming a concentration that may indicate the proximity of the anthill superficial sector.

Possible builders: The observations coincide with the data given by Bonetto (1959), Goncalves (1961) and Weber (1972); for that reason, the nest is recognized as belonging to the species Acromyrmex (Acromyrmex) close to A. ambiguus Emery.

(a)   (b)

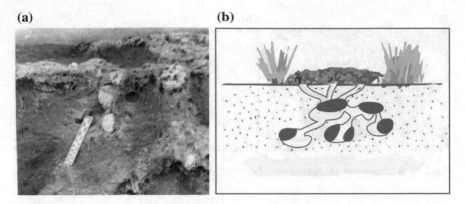

Fig. 2.14 a Ichnogenus aff. Acromyrmex (Acromyrmex) close to A. ambiguus Emery; **b** Scheme

(a)   (b)

Fig. 2.15 a Ichnogenus aff. Acromyrmex (Moellerius) striatus Roger; **b** Scheme

Ichnogenus Aff Acromyrmex (Moellerius) Striatus (Roger) (Fig. 2.15a, b)

The finding comes from the Centinela del Mar locality, General Alvarado County, Buenos Aires Province—Pleistocene, Bonaerian stage—Layer "C" (Tonni et al. 1987, Laza 1997). The preserved structure corresponds to two practically full chamber molds that converge on the filler of a communication duct. Both chambers, of elliptical contour, are about 150 mm long by 60 mm in diameter. Despite the erosion, the inner surface of the polished walls is observed in chambers and communication ducts. The filler is composed of the same rock bearing, cemented by carbonate. A great variety of root molds was found associated with the anthill.

Possible builders: The morphology of the discovered remains coincides with the descriptions made by Bruch (1916), De Santis (1941), Carbonell-Mas (1943), Bonetto (1959) and Goncalves (1961); the material was assigned to a nest of the species Acromyrmex (Moellerius) aff. M. (M.) striatus (Roger).

(a)   **(b)**

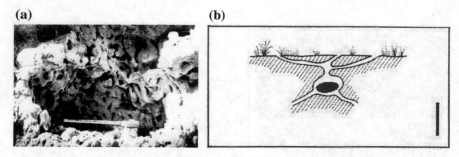

Fig. 2.16 a Ichnogenus aff. Acromyrmex (Acromyrmex) lundi Guerin **b** Scheme

Record in paleosols of Pampasia: A second specimen corresponding to the same ichnospecies was found in the longest section of the coastal cliffs from Coronel de Marina Leonardo Rosales County, Buenos Aires Province. Upper levels of the Pliocene Monte Hermoso Formation, Chapalmalalan stage. It was only possible to take photos and measurements of the mentioned specimen.

Ichnogenus Aff. Acromyrmex (Acromyrmex) Lundi Guerin (Fig. 2.16a, b)

This finding took place close to the Río Quequén Grande banks, near the archeological site Zanjón Seco (38° 10′ S–59° 10′ W), Necochea County, Buenos Aires Province. Pleistocene, Guerrero Member of the Luján Formation, Lujanian stage. It was possible only to take photos and measurements of the mentioned specimen (Laza 1997).

The erosion uncovered the central pot and the nearby zone. The general shape of the structure is framed in a trapezoid; the smallest base is 0.60 m and the largest 0.95 m; the height is 0.60 m. The largest part of this figure is irregularly occupied by the chamber. This shows the inner wall with the duct mouths ending into it, many of them tangentially. These features confer a mamelon or pad-like appearance to the chamber inner wall. The diameter of ducts that flow into the chamber is constant, of 10 mm, except for some globular expansions that reach up to 30 mm. The structure is built in silt-clayish sediments with a diagenesis level greater than that of a passive filling—silty-sandy-, which the erosion eliminated.

Possible builders: The general characteristics of the nest coincide with the data given by Gallardo (1916), De Santis (1941), Bonetto (1959) and Weber (1972). The material was assigned to a nest of the species Acromyrmex (Acromyrmex) aff. A. (A.) lundi Guerin.

Ichnogenus Aff. Trachymyrmex Emery (Fig. 2.17)

Frenguelli (1938b) described this fossil; he attributed the spheres to scarabaeinae Phanaeus nests and their gigantism as counterpart to the megafauna of that time

Fig. 2.17 Ichnogenus aff.
Trachymyrmex Emery

20 cm

(Lujanian stage). The examination carried out by the present author dismisses such assignation and attributes the specimens as fillings of both chambers or as fungus-growing ant anthills belonging to the Attini tribe. The description by Frenguelli (1938b) contributes to such assignation: "Its diameter is 82 and 87 mm, respectively, with walls reaching a thickness up to 20 mm and its inlet hole around 16 mm in diameter. The surface and walls formed by greenish grey hardened silt underwent the same diagenetic processes that transformed locally all the sediment in a mass of root pellets irregularly interwoven, in a complex tangle where non concretionary slit portions were trapped in the mesh. These portions, when easily destroyed by erosion and deflation, left rough and cavernous surfaces". The shape and size of the pieces remind us of the fungus-growing ants of Trachymyrmex Emery, a genus to which they could be doubtfully attributed.

Ichnogenus Aff. Pheidole (Fig. 2.18)

The site where the material was found is placed 8 km SE of the city of Santa Rosa, on the route linking that city with the Cerros Colorados locality, La Pampa Province, Miocene, Cerro Azul Formation, Huayquerian stage.

The fragment measurements, totally impregnated by "tosca", are the following: 230 × 170 × 135 mm. It has a system of long and short ducts. Many of the short ones of 4 mm end in "cul-de-sac," while the long ones of 6 mm are scarce. Despite the elaboration of the nest mass, the chambers are relatively scarce; one chamber with a greater tubular form of roughly 65 mm long by 50 mm wide stands out; and the long and short ducts flow into it. The other chambers, also in irregular tubular shape, measuring about 30 × 12 mm, have a limited number of long and short entrances. It was observed that the inner surface of some sectors in the largest chamber and all

Fig. 2.18 Ichnogenus aff.
Pheidole actual

the smallest chambers show significant isolation. Inside the largest chamber remains of ants and their cocoons were found, in addition to several dermal plaques of a Squamata, forming part of the filling material.

The 18 cocoons have the morphology assigned to Attini by Wheeler (1910); many of them have sediment incrustations, including a specimen with remains of an ant attached. The extreme of the complete pieces has a sharp end, though most of them have an open end, sign of a possible eclosion. The color of most pieces is light brown, whereas five of the best preserved cocoons are yellowish and the surface shows a silky texture. Their measurements: 5.2–4 mm long; 2.5–3 mm in diameter.

Possible builders: Among the sediments found within the largest chamber, fragmented remains of ants corresponding to the Attini tribe were found. Such fragments correspond to various individuals: sternites and tergites, one coxa and two small pronota, one of them well preserved. The pronotum has diagnostic features, such as the sculptured surface and its thorns; these are identified by their development, orientation and shape. The material presents lateral pronotal thorns more developed, leaning forward; rear mesonotal thorns hardly developed and the lower prenotal ones hardly visible, developed on a bulge. According to observations, illustrations and published diagnosis (Bonetto 1959; Weber 1972; Kusnezov 1978), it is concluded that the found remains correspond to an ant construction related with the genus Pheidole maybe developing shapes adapted to mesophyll environments at that moment.

Associated fauna: next to the ant remains and their cocoons (some of them embedded by "tosca") several dermal scutes, of a Squamata reptile were found. The nests of Attini ants, whose one of the representatives is the ichnofossil under study, usually become sites for oviposition of several reptiles. Vaz Ferreira et al. (1970, 1973) and Ferreira Brandao and Vanzolini (1985) mentioned plentiful examples related to reptile inquilinism in anthills of several species of this tribe.

The 21 scutes are long in shape, irregular, with rounded edges (3 mm long by 2 mm wide). Most of them present a notorious apex with both surfaces granular, light gray; two of them differ in color (dark brown) and in morphology, presenting mamelons in the center of both faces.

Ichnogenus Aff. Pheidole Spininodis Forel (Fig. 2.19)

The material comes from the coastal cliffs to the west of the city of Miramar, General Alvarado County, Buenos Aires Province. Pliocene-Pleistocene, Miramar Formation, Ensenadan stage. It was only possible to photograph and measure this specimen.

The set of chambers and ducts is framed in a vertical surface of 250 mm high by 200 mm wide. The central anthill duct, with a constant diameter of 8 mm, is arranged vertically (with some bends) and the four chambers, horizontally. The first one—upper—is shown with a widening of 30 mm in diameter toward both sides of the central duct; 60 mm underneath there are two chambers at both sides of the central duct of 60 mm long by 20 mm in diameter each one. In the deepest zone, there is a chamber of about 100 mm long by 40 mm wide joining the communication duct, getting narrower until ending in sharp points toward the opposite extreme. The nest was found in a sedimentary package of 450 mm thick, limited in its lower part by a thin and eminent "tosca" layer.

Possible builders: The reviewed structure coincides with Bruch's description and figures assigned to anthilla of Pheidole spininodis Forel (1916).

Ichnogenus Aff. Pogonomyrmex Bruchi Forel (Fig. 2.20)

The original place of the finding is at 400 m to the west of Punta Hermengo, General Alvarado County, Buenos Aires Province. Pliocene-Pleistocene, Miramar Formation, Ensenadan stage. The nest shows two globular chambers of 50 and 70 mm in diameter, and at the same level, other two eroded of the same size. Next to them, there appear overlapped chambers (at different levels) which are vertically long or short ducts of about 60 mm long by 5 mm in diameter. Cuts of several ducts of similar size distributed in the area can be observed. Possible builders: The fossil structure is

Fig. 2.19 Ichnogenus aff. Pheidole spininodis Forel

(a)   **(b)**

Fig. 2.20 Ichnogenus aff. Pogonomyrmex coarctatus Forel

similar to the nest description of Pogonomyrmex bruchi Forel. The species founding was performed by Bruch (1916) who described the nest.

See also Gallardo (1932) and Kusnezov (1963).

Fig. 2.21 Ichnogenus aff. Forelius chalybaeus Emery (**a**, **b** fossils; **c**, present)

Record in paleosols of Pampasia: Two structures similar in shape and dimension to those previously described were observed in the Monte Hermoso cliffs, at levels corresponding to the Pliocene, Chapalmalalan stage.

Ichnogenus Aff. Forelius Chalybaeus Emery (Fig. 2.21a, b and c)

This ichnofossil was found in Santa Clara del Mar, General Pueyrredón County, Buenos Aires Province. Pleistocene, Guerrero Member of the Luján Formation, Lujanian stage. It was only possible to photograph and measure the specimen.

In a vertical cross section, it may be seen that the nest is developed from a central duct, toward both sides. The chambers follow one another horizontally in depth; elongated (50–80 mm) and narrow (15 mm), linked by a series of secondary ducts almost vertical of 10 mm in diameter that connect the chambers with one another. The nest lateral extension decreases in depth, presenting, in a cross section, a triangular figure whose base is placed at its entrance.

Possible builders: The species Forelius chalybaeus Emery, described and featured by Bruch (1916), is recognized as the builder of the described nest.

Record in paleosols of Pampasia: Two nests similar to the one described were found close to the fishing breakwater in Punta Hermengo, General Alvarado County, in a paleosoil of the same age.

Other Mentions of Anthills in Pampasia

Santa Fe Province, Belgrano County. Pleistocene, Tezanos Pinto Formation. Lujanian stage.

Iriondo and Krohling (1996) indicated "numerous fossil anthills of 2–3 m long and 0.50–2 m high of varied shapes, filled with more consolidated silt, with black "patinas" in the middle portion of the profile. Santa Fe and Buenos Aires Provinces, between Rosario and Campana localities, Río Paraná gullies.

Pleistocene, Bonaerian stage. Voglino (1999) mentions and attributes activity signs to termites or ants, with nests up to 0.50 m long and ducts of 2–3 mm in diameter.

2.3.3.2 Termite Nests

Ichnogenus Tacuruichnus Farinai Genise 1997 (Fig. 2.22a, b)

The impossibility of removing this kind of nest from the paleosol (discussed in Genise and Bown 1994b) limits the holotype designation to the selection of a photo-type as representative of the type specimen (Genise 1997), expression applied to other specimens mentioned in this work. The material comes from the coastal cliffs of the Terrazas del Marquesado beach, 1000 m south of Punta Vorohue, General Alvarado County, Buenos Aires Province. Pliocene, Marplatan (Barrancalobian) stage. A section of this trace fossil, cup-shaped, was exposed by natural erosion in the costal cliffs when it was discovered. It is assumed that originally this trace fossil was completed with a sub-aerial mound from which only the basal part, cup-shaped, was preserved in the paleosol, whereas the superficial structure was eroded. The preserved cup was recently exposed laterally by marine erosion, revealing its current shape, its front part lost, and its rear side preserved within the cliff. The diameter of the exposed arch is 80 cm, while the wall is 10 cm wide, and it is perforated by a system of anastomosed tunnels of different diameters between 1 and 7 mm. In the insect traces, as usual, the wall is stronger than the hosting paleosoil and therefore

(a)   (b)

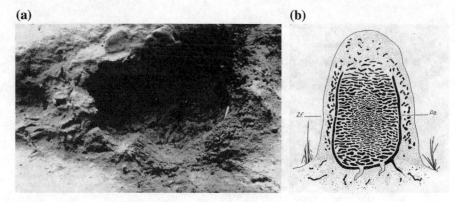

Fig. 2.22 **a** Ichnogenus Tacuruichnus farinai Genise 1997; **b** scheme

it is in relief. The inner part of the cup does not preserve any activity signal and is currently filled with passive sediments. Outside the wall, a system of peripheral galleries can be seen in the cliff at both sides of the cup. The exposed part of this system consists of 2–4 main tunnels in each side, of 10–15 cm long and 1–2 cm wide connected to others of only 1 or 2 mm in diameter. It is possible to observe remains of the tunnels even beyond 30 cm from the wall.

Possible builders: The described architecture is similar to the hypogeous sector of the nest of Cornitermes cumulans Kollar (Nasutitermitinae) (Genise 1997).

See descriptions and figures by Emerson (1952) and Grassé (1984) regarding nests of the aforementioned species.

Record in paleosols of Pampasia:

(a) In paleosols of the lower section of the Atlantic Ocean, coastal outcrops in the city of Mar del Sur, General Alvarado County, Buenos Aires Province, abundant specimens of Tacuruichnus were found. Such levels were attributed to the Middle Pleistocene, Bonaerian stage, Ctenomys kraglievichi Biozone (Eduardo Tonni and Esteban Soibelzon, personal communication).

(b) In 2016, the finding of termite nests in Pleistocene sediments of the Arroyo Toropí, in the province of Corrientes was published by Erra et al. The edited photographs show several Tacuruichnus specimens.

Ichnogenus Barberichnus Bonaerensis Laza 2006a (Fig. 2.23a, b)

The materials corresponding to the holotype were found in excavations carried out in the city of La Plata (51St. Avenue, between the 9th. and 10th. Streets), Buenos Aires Province, in the Pleistocene, lower section of Buenos Aires Formation, Bonaerian stage, Ctenomys kraglievichi Biozone. The two fragments that form part of the holotype belong to the only collected termite nest. The fragments, which correspond to the hypogeous part of the nest, were found very close one another, without totally completing the structure but representing most of it. One piece is 29 × 27 cm, and

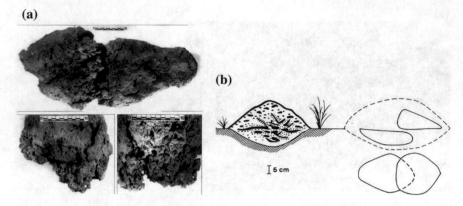

Fig. 2.23 **a** Ichnogenus Barberichnus bonaerensis Laza 2006a; **b** scheme

the other is 37 × 26 cm, and they can be aligned on a horizontal axis suggesting an elongated shape, of around 25 cm high. A system of ducts is observed: (a) longer of oblong section of 13 mm high × 15 mm wide arranged without any apparent order; (b) shorter, of circular section of 1–3 mm. The confluence of several longer ducts causes hollows that are interpreted as chambers—highly irregular- some of 30 × 45 mm. In between the longest ducts, the system of shorter ducts is developed, being highly abundant in some sectors. In the central zone of the termite nest as well as inside some dissected ducts, it can be observed a sharp change of color and texture of the rock. The sediment is light brown with scattered specks of calcium carbonate of 10 mm in diameter, whereas a violet brown halo surrounds the duct and chamber systems. The inner surfaces of the tunnels and chambers appear covered by a patina of waxy appearance. The texture of these surfaces is quite polished, differently from the sediment bearing which is more porous. Possible builders: The general characteristics of these termite nests allow relating them to members of the Termitinae subfamily, with representatives of Amitermes and Termes genuses as possible builders.

Record in paleosols of Pampasia: (a) Sector of coastal cliffs of Punta Negra, Necochea County, Buenos Aires Province. Pliocene, Sanandresian stage. (b) Sector of cliffs of Barranca de Los Lobos, General Pueyrredón County, Buenos Aires Province. Pliocene, Vorohuean stage.

Ichnogenus Aff. Procornitermes Emerson (Fig. 2.24a, b, and c)

The material comes from the gullies placed between Playa Las Palomas and Barranca de Los Lobos, General Pueyrredón County, Buenos Aires Province. Pliocene, Chapalmalal Formation. Chapalmalalan Stage (Banco VIII de Kraglievich 1952) (Laza and Tonni 2004). At levels close to those of the finding, other two specimens were observed, but it was impossible to have access. The piece is sagittally cut oriented to the longest axis (vertical), lacking a small portion in one end. The general

(a) (b) (c)

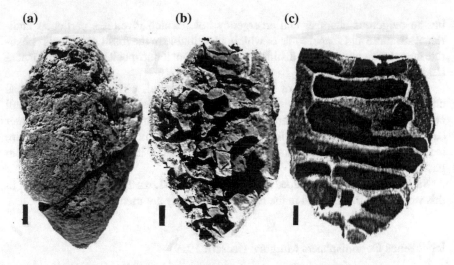

Fig. 2.24 Ichnogenus aff. Procornitermes Emerson (**a**, **b** fossil; **c**, present)

shape corresponds to that of a pine cone with globosity on one side. Within the cut, there are partitions developing and anastomosing which follow directions according to the shortest axis of the piece. The thickness of the walls and inner partitions is constant, between 7 and 10 mm. In between the partitions, there are fine separations that are believed to have been larger at another time, obliterated by "tosca" accretion. In the center of the body, it is observed a partition that separates it medially oriented to the longest axis, whereas an end shows a hollow interpreted as an entry point to the nest. The general dimension is 115 mm high and 70 mm wide.

Possible builders: The dimensions as well as the observations on the present material, descriptions and figures of the known species of the genus allow asserting that the nest belongs to Procornitermes Emerson (Snyder 1949) genus. See also Silvestri (1903), Emerson (1952) and Grassé (1984).

Ichnofamily Coprinisphaeridae Genise 2004

Coprinisphaera is one of the most common trace fossils in paleosols of the Cenozoic of South America and one of the first being recorded (Frenguelli 1938a, Roselli 1939). It has been cited for different localities and ages of the Cenozoic in South America, Europe, Asia, Antarctica and Africa (Genise et al. 2000; Krell, 2000 and references; Laza 2006b; Sánchez 2009). These traces correspond to beetle breeding of the Scarabaeinae subfamily, when isolated to the ichnogenus Coprinisphaera Sauer (1955) and when grouped to the ichnogenus Quirogaichnus (Laza 2006). Each ball has an ovigerous chamber and emergency hole, which gives it a particular additional structure and shape, some of which, identifiable in the fossil specimens constitute ichnotaxobases used for distinguishing different ichnospecies of Coprinisphaera (Laza 2006; Sánchez 2009).

The fossil trace consists of spherical, sub-spherical, piriform and bispherical chambers, formed by a wall with a discreet emergency hole. They can show a small chamber linked to this hole through a corridor and connected outside. The inner cavities have mostly passive filling, being empty in some cases. The chambers are isolated in contact with the matrix or situated in one of its cavities, normally with passive filling.

Such structures were grouped in six ichnospecies (Laza 2006; Sánchez 2009). In this work, only those found in the region under study are mentioned.

Ichnogenus Coprinisphaera Murguiai (Roselli 1939)

It is formed by isolated chambers, spherical and sub-spherical, with a wall quite thick. In one of the poles, the chamber wall is completely perforated by a medium sized hole (around ¼ in equatorial diameter), which has a cylindrical contour in longitudinal section. There are no remains of a secondary chamber or of additional structures around the hole. Inside the chamber there is, as a rule, passive infilling (Sánchez 2009). The type material comes from the Paleocene–Eocene, the Asencio Formation, Uruguay.

The examined specimens show great size variation. Coprinisphaera murguiai reminds the breeding balls built by dung beetles (Scarabaeinae) of Coprini tribe, Dichotomiina sub-tribe (diggers) included in Pattern II of nidification by Halffter and Edmonds (1982) and of Scarabaeini tribe, Canthonina sub-tribe (wheelers) included in Pattern III by the mentioned authors. Assignation based on the isolation condition, spherical shape, wall thickness, possible location of the ovigerous chamber within the provision chamber and current geographic distribution of the Subfamily.

Record in paleosols of Pampasia:

Miocene, Cerro Azul Formation, Huayquerian stage. Arroyo Venado, between Guaminí and Carhué localities, Buenos Aires Province. Provincial Route, 152.2 km from the General Acha locality, from Route 35, La Pampa Province. Gullies to the north of Salinas Grandes de Hidalgo, La Pampa Province. Miocene? Cerro Azul Formation? (Huayquerian stage?). Southern Córdoba Province, in an outcropping paleosoil in interdune valleys.

Miocene, Paso de Las Carretas Formation, Huayquerian-Montehermosan stages. Materials mentioned by Guiñazú (1960) who wrongly attributed the nests to the dynastine Diloboderus. Coronel Pringles Department, San Luis Province. In Paso de Las Carretas Formation, Santa Cruz (1979) mentioned the finding of "vespids and coleoptera nests" in the rivers Quinto and Conlara basins, in San Luis Province.

Miocene, Las Mulitas Formation. Huayquerian-Montehermosan stages. Río Los Chorrillos, tributary of the Río Quinto, San Luis Province. González (1979) mentioned "insect nests".

Pliocene, Irene Formation. Chapalmalalan stage. Río Quequén Salado, Buenos Aires Province.

Pliocene? Monte Hermoso Formation? (Montehermosan stage?), specimens from an excavation at 10–12 m depth in the Hansen farm, between El Zorro and Guisasola railway stations, Coronel Dorrego County, Buenos Aires Province.

Pliocene, Irene Formation, Chapalmalalan stage. Río Quequén Salado near National Route 3, Buenos Aires Province. Several specimens (Aramayo et al. 2004a, b).

Pliocene, Chapalmalal Formation, Chapalmalalan stage. Monte Hermoso cliffs, Coronel de Marina Leonardo Rosales County, Buenos Aires Province.

Pliocene, Chapalmalal Formation, Chapalmalalan stage. Bajada Martínez de Hoz, General Pueyrredón County, Buenos Aires Province. Fifteen small specimens (Tetraechma or next canthonini) found inside caves with Actenomys, close to their passive filling.

Pliocene, San Andrés Formation, Sanandresian stage (Layer "A" of Tonni et al. 1996). Necochea County coastal cliffs, Buenos Aires Province.

Pliocene, Barranca de Los Lobos Formation, Barrancalobian stage. Between Fortín 88 and Las Palomas Beach, General Pueyrredón County, Buenos Aires Province. Bajada Martínez de Hoz, General Pueyrredón County, Buenos Aires Province.

Pliocene, San Andrés Formation, Sanandresian stage. At 100 m to the north of the Chapalmalal Hotels, General Pueyrredón County, Buenos Aires Province. Small nest (Canthonino?) associated to a Conepatus sp. skull.

(b) Punta Negra, 11 km to the west of Necochea (level "A" of Tonni et al. 1996).

Pleistocene, Buenos Aires Formation, Bonaerian stage. Baradero County, Buenos Aires Province. Centinela del Mar Beach, General Alvarado County. Buenos Aires Province.

Pleistocene, Arroyo Seco Formation, Bonaerian stage. Santa Isabel Beach, General Pueyrredón County, Buenos Aires Province.

Pleistocene, Guerrero Member of the Luján Formation, Lujanian stage. Río Salado, Las Flores County, Buenos Aires Province, associated to a Stegomastodon molar.

Pleistocene, Luján Formation, (Guerrero Member), Lujanian stage. Lake Guaminí, Stream El Venado, Buenos Aires Province.

   Pleistocene-Holocene, La Postrera Formation, Lujanian stage, Tres Arroyos County, Buenos Aires Province. Level "S" of the archeological site Arroyo Seco I (Fidalgo et al. 1986); several specimens.

Possible remains of Coprinisphaera.

(a) Pleistocene, Tezanos Pinto Formation, Lujanian Stage. Tortugas Locality, Belgrano Department, Santa Fe Province. Iriondo and Krohling (2001) pointed out "nodules that are interpreted as a result of the activity of dung beetles".
(b) Pleistocene, Levels 3 and 4 (Voglino 1999), Río Paraná gullies between the cities of Rosario and Campana.
(c) Late Pleistocene. Coprinisphaera? San Pedro and Baradero cities. Buenos Aires Province (Nabel 1993).
(d) Pleistocene, Toropí and Yupoí Formations, Río Paraná gullies, Empedrado City, Corrientes Province (Lutz and Gallego 2001).

Ichnogenus Coprinisphaera Isp "A" Sánchez 2009 (Fig. 2.25)

Isolated chambers, spherical and sub-spherical, with a quite thick built wall. In one of the poles, the chamber wall includes a cavity or secondary chamber, hemispherical and small, open to the outside. A very narrow passageway on the chamber floor connects it with the main chamber. Within this chamber, there is passive filling as a rule.

The examined specimens show great size variation.

Coprinisphaera isp "A" reminds the breeding balls built by dung beetles (Scarabaeinae) of Coprini tribe, Dichotomiina sub-tribe (diggers) included in Pattern II of nidification by Halffter and Edmonds (1982). Assignation based on the isolation condition, spherical shape, wall thickness, possible location of the ovigerous chamber within the provision chamber and current geographic distribution of the Subfamily.

   Record in paleosols of Pampasia:

(a) Pliocene, Irene Formation, Chapalmalalan stage. Río Quequén Salado and National Route 3, Buenos Aires Province. Several specimens (Aramayo et al. 2004a, b).
(b) Luján Formation, Guerrero Member. Stream El Venado, Guaminí County, Buenos Aires Province.

Ichnogenus Coprinisphaera Akatanka Cantil et al. 2013 (Fig. 2.26)

Isolated chambers, bispherical, with a relatively thin built wall. Consisting of a spherical cavity or main chamber connected by a passageway to another secondary one which is smaller and also spherical. Externally, the wall of this structure has a very sharp bottleneck which defines both chambers. Both chambers, as a rule, have passive filling inside. It is the first ichnoshape attributed to a ghoul beetle.

Fig. 2.25 Ichnogenus
Coprinisphaera "A" Sánchez
2009

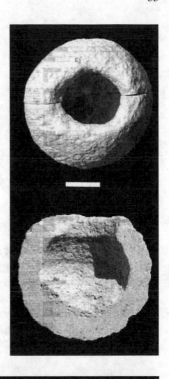

Fig. 2.26 Ichnogenus
Coprinisphaera akatanca
Cantil et al. 2013

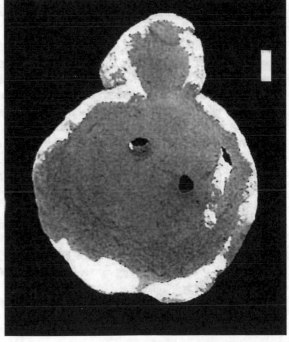

Fig. 2.27 a Ichnogenus
Quirogaichnus coniunctus
Laza 2006b (a, fossil); b
Scheme)

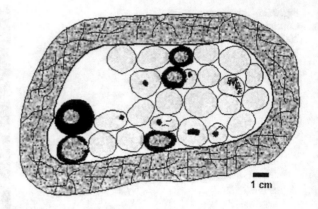

C. akatanka reminds the breeding balls built by beetles (Scarabaeinae) of Coprini tribe, Phanaeina sub-tribe (diggers) of Pattern II of Halffter and Edmonds (1982) and Scarabaeini tribe, Canthonina sub-tribe (wheelers) corresponding to Pattern V of the same authors. Assignation based on the isolation condition, bispherical shape of the ovigerous chamber and current geographic distribution of such beetles.

Record in paleosols of Pampasia: San Andrés Formation, Sanandresian stage (Layer "A" of Tonni et al. 1996). Necochea cliffs, Buenos Aires Province.

Ichnogenus Quirogaichnus Coniunctus Laza 2006b (Fig. 2.27a, b)

The grouping condition of the specimens in a cavity is the main characteristic for separating this ichnogenus from Coprinisphaera Sauer. The specimen grouping is an ichnotaxobase of generic value in other groups of insect trace fossils in paleosols (Genise 2000, 2004). The modern counterpart of this ichnogenus is the type of compound nest of dung beetles defined by Halffter and Edmonds (1982, p. 33) for some Neotropical Canthonina. The materials consist of spherical to sub-spherical chambers grouped in a larger cavity. Such chambers are built by a thin wall and a cylindrical emergency hole. The holotype is formed by 25 spheres from 12 to 22 mm in diameter grouped in a cavity of 125 mm long by 70 mm high; the thickness of the sphere walls is from 2 to 3 mm. Some spheres have holes from 2.7 to 3.4 mm in diameter. Passive filling.

Quirogaichnus coniunctus is similar to compound nests built by beetles of the Scarabaeinae Subfamily, Scarabaeini tribe, Canthonina sub-tribe (wheelers), included in pattern V of nidification of Halffter and Edmonds (1982). The assignation is based on the grouping conditions of the spheres and on the geographic distribution of the subfamily.

Record in paleosols of Pampasia.

Miocene, Cerro Azul Formation, Huayquerian stage. Holotype found in sections of National Route 33 at its intersection with Provincial Route 85, near Guaminí County, Buenos Aires Province.

Miocene, Cerro Azul Formation, Huayquerian stage. Gullies to the south of Provincial Route 152, 2 km. to the east of the town of General Acha, Utracán Department, La Pampa Province. A specimen with its chamber and four nest-balls inside.

Pleistocene, Ensenadan stage. Sitio Los Galpones, Saltos del Guaviyú, Entre Ríos Province. It consists of 10 spherical chambers of 11–27 mm in diameter, without visible holes and thin walls of 4 mm. They were found by J. Frenguelli in 1920 grouped under a glyptodont (Neosclerocalyptus) shell and mentioned then by him in his paper of 1938a.

References

Abel O (1935) Vorzeitliche Lebensspuren. Gustav Fischer, Iena
Aceñolaza F, Aceñolaza G (2004) Trazas fósiles en unidades estratigráficas del Neógeno de Entre Ríos. E: Temas de la biodiversidad del Litoral fluvial argentino. INSUGEO, Miscelánea 12:19–24
Ameghino F (1880) La antigüedad del hombre en el Plata. Ed. La Cultura Argentina, II, 378. Buenos Aires
Ameghino F (1908) Las formaciones sedimentarias de la región litoral de Mar del Plata y Chapalmalán. Anales Museo Nac Historia Nat 3(10):343–428
Andreis R (1972) Paleosuelos de la Formación Musters (Eoceno Medio), Laguna del Mate, Provincia de Chubut, República Argentina. Rev Asoc Argent de Mineralogía, Petrología y Sedimentología 3:91–97
Andreis R (1981) Identificación e importancia geológica de los paleosuelos. Ed. Universidad UFRGS. Libro-Texto 2. Porto Alegre, pp 67
Aramayo S (1999) Nuevo registro de icnitas en la Formación Río Negro (Mioceno tardío -Plioceno temprano), Provincia de Río Negro, Argentina. XII Jornadas Argentinas de Paleontología de Vertebrados. Resúmenes: 3. La Plata and Luján
Aramayo S, Manera de Bianco T (1987a) Hallazgo de una icnofauna continental (Pleistoceno tardío) en la localidad de Pehuén-Có (Partido de Coronel Rosales, provincia de Buenos Aires. Argentina. Parte I: Edentata, Litopterna, Proboscidea. IV Congreso Latinoamericano de Paleontología. Boliv Actas I:516–531
Aramayo S, Manera de Bianco T (1987b) Hallazgo de una icnofauna continental (Pleistoceno tardío) en la localidad de Pehuén-Có (partido de Coronel Rosales), provincia de Buenos Aires, Argentina. Parte II: Carnivora, Artiodactyla and Aves. IV Congreso Latinoamericano de Paleontología. Boliv Actas I:532–547
Aramayo S, Manera de Bianco T (1996) Edad y nuevos hallazgos de icnitas de mamíferos y aves en el yacimiento paleoicnológico de Pehuén-Có (Pleistoceno tardío), provincia de Buenos Aires, Argentina. Asociación Paleontológica Argentina, Publicación Especial N° 4. 1ª Reunión Argentina de Icnología: 47–57
Aramayo S, Manera de Bianco T (1998) Primer registro de Caviidae (Rodentia) y Ursidae (Carnívora) en el yacimiento paleoicnológico de Pehuén-Có (Pleistoceno tardío), provincia de Buenos Aires, Argentina. 3ª Reunión Argentina de Icnología y Primera Reunión de Icnología del Mercosur. Resúmenes: 7

Aramayo S, Manera de Bianco T (2000) Primer hallazgo de icnitas de mamíferos pleistocenos en "Playa del Barco", Pehuén-Có, Provincia de Buenos Aires, Argentina. XVI Jorn de Paleontología de Vertebrados, San Luis. Actas: 5

Aramayo S, Manera de Bianco T (2009) Late Quaternary paleichnological sites from the Southern Atlantic Coast of Buenos Aires Province, Argentina: Mammal, bird and hominid evidence. Ichnos 15(1–2):25–32

Aramayo S, Manera de Bianco T, Bocanegra L (2003) Presencia de Taenidium Heer 1877 en el yacimiento paleoicnológico de Pehuen-Có (Pleistoceno tardío), provincia de Buenos Aires, Argentina. IV Reunión Argentina de Icnología and II Reunión de Icnología del Mercosur. Asoc Paleontológica Argentina. Publicación Especial 9:49–52

Aramayo S, Schillizzi R, Gutiérrez Téllez B (2004a) Coprinisphaera isp at the Irene Formation (Early to Middle Pliocene), Quequén Salado River, Buenos Aires province, Argentina. First International Congress on Ichnology. Abstract book: 13. Trelew. Argentina

Aramayo S, Barros M, Candel S, Vecchi L (2004b) Mammal and bird footprints at Río Negro Formation (Late Miocene-Early Pliocene), Río Negro Province, Argentina. First International Congress on Ichnology, Abstract book: 14. Trelew. Argentina

Aramayo S, Gutiérrez Téllez B, Schillizzi R (2005) Sedimentologic and paleontologic study of the southeast coast of Buenos Aires province, Argentina: A Late Pleistocene Holocene paleoenvironmental reconstruction. J S Am Earth Sci 20:65–71

Aramayo S, Di Martino V, Sánchez N (2007) Nuevo registro de huellas de mamíferos pleistocenos en la localidad de Monte Hermoso, provincia de Buenos Aires, Argentina. XXIII Jornadas Argentinas de Paleontología de Vertebrados. Resúmenes: 2. Trelew, Argentina

Bedatou E, Melchor R, Bellosi E, Genise J (2008) Crayfish burrows from Late Jurassic- Late Cretaceous continental deposits of Patagonia: Argentina. Their palaeoecological, palaeoclimatic and palaeobiogeographical significance. Palaeogeogr Palaeoclimatol Palaeoecol 257:169–184

Beilinson E (2010) Icnological and paleoclimatic analysis of weakly developed paleosols: Punta San Andrés Alloformation (Plio-Pleistocene, Buenos Aires province, Argentina). In: 18th international sedimentological congress, Mendoza, Argentina. Abstracts: 167

Bertling M, Braddy S, Bromley R, Demathieu GG, Genise J, Mikulas R, Nielsen J, Nielsen K, Rindsberg A, Schlirf M, Uchman A (2006). Names for trace fossils: a uniform approach. Lethaia 39:265–286

Bonetto A (1959) Las hormigas cortadoras de la provincia de Santa Fe (Géneros Atta y Acromyrmex). Dirección General Recursos Naturales, Provincia de Santa Fé, p 77

Bown T (1982) Ichnofossils and rhizoliths of the nearshore fluvial Jebel Qatrani Formation (Oligocene), Fayum Province, Egypt. Palaeogeogr Palaeoclimatol Palaeoecol 40:255–309

Bown T, Kraus M (1983) Ichnofossils of the alluvial Willwood Formation (Lower Eocene), Bighorn Basin, Northwestern Wyoming, U.S.A. Palaeogeogr Palaeoclimatol Palaeoecol 43:95–128

Bown T, Laza J (1990) A Miocene fossil termite nest from southern Argentina and its paleoclimatological implications. Ichnos 1:73–79

Bown T, Hasiotis S, Genise J, Maldonado F, Brouwers E (1997) Trace fossils of Hymenoptera and other insects and paleoenvironments of the Claron Formation (Paleocene and Eocene), Southwestern Utah. U.S. Geol Surv Bull 2153:42–58

Bravard A (1857a) Observaciones geológicas sobre diferentes terrenos de transporte en la hoya del Plata. Library of the La Prensa newspaper. Buenos Aires

Britt B, Scheetz R, Dangerfield A (2008) A suit of dermestid beetle traces on dinosaur bone from the Upper Jurassic Morrison Formation, Wyoming, U.S.A. Ichnos, 15: 59-71. Bromley, R and Asgaard, U (1979). Triassic freshwater ichnocoenoses from Carlsberg Fjord, East Greenland. Palaeogeogr Palaeoclimatol Palaeoecol 28:39–80

Bromley R (1990) Trace fossils. Bilogy and taphonomy. Unwin Hyman. London, p 280

Bromley R, Ekdale A (1998) Ophiomorpa irregulaire (trace fossil): redescription from the Cretaceous of the Book Cliffs and Wasatch Plateau, Utah. J Paleontol 72(4):773–778

Brown R (1934) Celliforma spirifer, the fossil larval chambers of mining bees. J Wash Acad Sci 24:532–539

Brown R (1935) Further notes on fossil larval chambers of mining bees. J Wash Acad Sci 25:526–538

Brown R (1941) The comb of a wasp nest from the Upper Cretaceous of Utah. Am J Sci 239:54–56

Bruch C (1916) Contribución al estudio de las hormigas de la provincia de San Luis. Rev del Museo de La Plata 23:291–357

Bruch C (1917) Costumbres y nidos de hormigas. I. Anales Soc Científica Argent 83:302–316

Buatois L, Mángano M (1995a) The paleoenvironmental and paleoecological significance of the Mermia ichnofacies: an archetipycal subaqueous nonmarine trace fossil assemblage. Ichnos 4:151–161

Buatois L, Mángano M (1998) Trace fossil analysis of lacustrine facies and basins. Palaeogeogr Palaeoclimatol Palaeoecol 140:367–382

Buatois L, Mángano M (2004) Animal-substrate interactions in freshwater environments: applications of ichnology in facies and sequence stratigraphic analysis of fluvio-lacustrine successions. In: McIlroy D (ed) The application of Ichnology to Palaeoenvironmental and Stratigraphic Analisys, vol 228. Geological Society London. Special Publications, pp 311–333

Buatois L, Mángano M (2007) Invertebrate ichnology of continental freshwater environments. In: Miller III W (ed) Trace fossils. concepts, problems, prospects. Elsevier, Amsterdam, pp 285–323

Buatois L, Mángano M, Aceñolaza F (2002) Trazas fósiles. Museo Paleontológico "Egidio Feruglio". Edición Especial N° 2. Trelew, Chubut, Argentina

Buchmann F, Pereira Lopes R, Caron F (2000) Icnofosseis (paleotocas e crotovinas) atribuidos a mamiferos extintos no sudeste e sul do Brasil. Rev Brasileira de Paleontol 12:247–256

Buffetaut E (2000) A forgotten episode in the history of dinosaur ichnology: Carl Degenhardt´s report on the first discovery of fossil footprints in South America (Colombia, 1839). Bol Soc Geol France 171(1):137–140

Cantil LR, Sánchez MV, Bellosi ES, González MG, Sarzetti LC, Genise JF (2013) Coprinisphaera akatanka isp. nov.: the first fossil brood ball attributable to necrophagous dung beetles associated with an Early Pleistocene environmental stress in the Pampean region (Argentina). Palaeogeogr Palaeoclimatol Palaeoecol 386:541–554

Carbonell-Mas C (1943). Las hormigas cortadoras del Uruguay. In: Revista Asociación Ingenieros Agrónomos, vol 3. Montevideo, pp 1–12

Casamiquela R (1974) El bipedismo de los megaterioideos. Estudio de pisadas fósiles en la Formación Río Negro típica. Ameghiniana 11(3):249–282

Casamiquela R (1983) Pisadas del Pleistoceno (¿superior?) del Balneario de Monte Hermoso, Buenos Aires. La confirmación del andar bipedal en los megaterioideos. Cuad del Inst Super "Juan XXIII" 4:5–21

Chimento N, Rey L (2008) Hallazgo de una feca fósil en el Pleistoceno Superior—Holoceno Inferior del Partido de General Guido, provincia de Buenos Aires. Rev del Museo Argentino de Cienc Naturales, n.s. 10(2):239–254. Buenos Aires

Cione A, Tonni E, San Cristóbal J (2002) A Middle-Pleistocene marine transgression in Central-eastern Argentina. CRP 19:16–18

Cosarinsky M, Genise J, Bellosi E (2004) Micromorphology of modern epigean nests and possible termite ichnofossils: a comparative analysis. In: First international congress on ichnology. Abstract book: 26. Trelew, Argentina

Cox G, Roig V (1986) The occurrence in Argentina of Mima mounds occupied by ctenomyd rodents. J Mammal 67:428–432

Cox G, Mills J, Ellis B (1992) Montículos tipo Mima, posiblemente originados por hormigas en Necochea, provincia de Buenos Aires, Argentina. Rev Chil de Historia Nat V 65(3):311–318

Darry D (1911) Damage done to skulls and bones by termites. Nature 86:245–246

De los Reyes L, Cenizo M, Soibelzon E (2006) Una notable tafocenosis en el interior de paleo-ocuevas de la "Formación" Chapadmalal (Plioceno medio). Evidencias paleobiológicas para Thylophorops chapadmalensis (¿Didelphidae, Didelphimorphia?). Ameghiniana. Suplemento Resúmenes 43(4):33–34

De Santis L (1941) Las principales hormigas dañinas de la Provincia de Buenos Aires. Ministerio de Obras Públicas, Buenos Aires, p 40

Deschamps C, Tonni E, Verzi D, Scillato-Yané G, Zavala C, Carlini A, Di Martino V (1998) Bioestratigrafía del Cenozoico superior continental en el área de Bahía Blanca, Provincia de Buenos Aires. V Jornadas Geológicas y Geofísicas Bonaerenses. Actas 1:49–57

Dondas A, Isla F, Carballido J (2009) Paleocaves exhumed from the Miramar Formation (Ensenadan Stage-age Pleistocene) Mar del Plata Argentina. Quat Int 210(1–2):44–50

Dubiel R, Hasiotis S (1994a) Integration of sedimentology, paleosols, and trace fossils for paleo-hydrologic and paleoclimatic interpretations in a Triassic tropical alluvial system. Geol Soc Am Abs Programs 6:A-502

Dubiel R, Hasiotis S (1994b) Paleosols and rhizofacies as indicators of climatic change and groundwater fluctuations: the Upper Triassic Chinle Formation. Geol Soc Am Meet Abstracts: 11–12

Duringer P, Brunet M, Cambefort Y, Likius A, Mackaye H, Schuster M, Vignaud P (2000) First discovery of fossil dung beetle brood balls and nest in the Chadian Pliocene Australopithecine levels. Lethaia 33:277–284

Duringer P, Schuster M, Genise J, Mackaye H, Vignaud P, Brunet M (2007) New termite trace fossils: galleries, nests and fungus combs from the Chad basin of Africa (Upper Miocene-Lower Pliocene). Palaeogeogr Palaeoclimatol Palaeoecol 251:323–353

Ekdale A, Bromley R, Pemberton S (1984) Ichnology, trace fossils in sedimentology and stratigraphy. Soc Ecomomic Paleontol Mineralogists. Short Course Notes N° 15

Ellorriaga E, Visconti L (2002) Crotovinas atribuibles a grandes mamíferos del Cenozoico de la Provincia de La Pampa. IX Reunión Argentina de Sedimentología. Córdoba. Resúmenes, p 63

Emerson A (1952) The neotropical genera Procornitermes and Cornitermes (Isoptera Termitidae). Bull Am Mus N.H. 99:475–540

Erra G, Osterrieth M, Zurita A, Lutz A, Laffont E, Coronel J, Francia A (2016) Primer registro de termiteros fósiles para el Pleistoceno tardío de la región mesopotámica (Argentina): implicancias paleoambientales. Acta Biol Colomb 21(1):63–72

Estrada A (1941) Contribución geológica para el conocimiento de la "cangagua" de la región interandina y del Cuaternario en general del Ecuador. Anales de la Univ Central 66(312):405–488. Quito, Ecuador

Ferreira Brandao C, Vanzolini P (1985) Notes on incubatory inquilinism between Squamata (Reptilia) and the neotropical fungus-growing ant genus Acromyrmex (Hymenoptera: Formicidae). Papéis Avulsos de Zoologia 36(3):31–36

Fidalgo F, Tonni E (1982) Sedimentos eólicos del Pleistoceno tardío y Reciente en el área interserrana bonaerense. VIII Congreso Geol Argentino, Actas III:33–39

Fidalgo F, Meo Guzmán L, Politis G, Salemme M, Tonni E (1986) Investigaciones arqueológicas en el Sitio 2 de Arroyo Seco (Pdo. de Tres Arroyos—Provincia de Buenos Aires—República Argentina). In: New evidence for the Pleistocene people of the Americas, A.L. Bryan, editor. Center for the study of Early Man. University of Maine at Orono, pp 221–269

Frenguelli J (1921) Los terrenos de la costa Atlántica en los alrededores de Miramar (Provincia de Buenos Aires) y sus correlaciones. Acad Nac de Ciencias de Córdoba 24:325–485

Frenguelli J (1928) Observaciones geológicas en la región costera sur de la Provincia de Buenos Aires. Universidad Nacional del Litoral. Anales Fac de Cienc de la Educación 3:101–130

Frenguelli J (1930) Apuntes de geología uruguaya. Bol del Inst de Geol y perforaciones (Uruguay)

Frenguelli J (1938b) Nidi fossili di Scarabeidi e vespidi. Bolletino della Soc Geol Ital 57:77–96

Frenguelli J (1938a) Bolas de escarabeidos y nidos de véspidos fósiles. Physis XII:348–352

Frenguelli J (1939b) Nidos fósiles de insectos en el Terciario del Neuquén y Río Negro. Notas del Museo de La Plata (Paleontologia) 4(18):379–402

Frey R, Curran H, Pemberton G (1984a) Tracemaking activities of crabs and their environmental significance: the ichnogenus Psilonichnus. J Paleontol 58:333–350

Frey R, Pemberton S, Fagestrom J (1984b) Morphological, ethological and environmental significance of the ichnogenera Scoyenia and Ancorichnus. J Paleontol 58:511–528

Frey R, Pemberton S, Saunders T (1990) Ichnofacies and bathymetry: passive relationship. J Paleontol 64:155–158

Gallardo A (1932) Las hormigas de la República Argentina. Subfamilia mirmicinas segunda sección eumyrmicinae—género pogonomyrmex mayr. Anales Museo Nac Historia Nat 37:89–169

Genise J (1989) Las cuevas con actenomys (Rodentia, Octodontidae) de la formación chapadmalal (Plioceno Superior) de Mar del Plata y Miramar (provincia de Buenos Aires). Ameghiniana 26(1–2):33–34

Genise J (1997) A fossil termite nest from the Marplatan stage-age (late Pleistocene) of Buenos Aires province, Argentina as paleoclimatic indicator. Palaeogeogr Palaeoclimatol Palaeoecol 136:139–144

Genise J (1999) Paleicnología de insectos. Rev Soc Entomológica Argent 58(1–2):104–116

Genise J (2000) The Ichnofamily Celliformidae for Celliforma and allied ichnogenera. Ichnos 7(4):267–282

Genise J (2004) Ichnotaxonomy and ichnostratigraphy of chambered trace fossils in palaeosols attributed to coleopterans, ants and termites. In: McIlroy (ed) The application of ichnology to palaeoenvironmental and stratigraphic analysis. Geological Society vol 228. Special Publications, London, pp 419–453

Genise J, Bown T (1994a) New Miocene scrabeid and hymenopterous nest and Early Miocene (Santacrucian) paleoenvironments, Patagonian Argentina. Ichnos 3:107–117

Genise J, Bown T (1994b) New trace fossils of termites (Insecta: Isoptera) from the late eocene-early miocene of egypt, and reconstruction of ancient isopteran social behavior. Ichnos 3:155–183

Genise J, Bown T (1996) Uruguay Roselli 1938 and Rosellichnus, new ichnogenus: Two ichnogenera for cluster of fossil bee cells. Ichnos 4:199–217

Genise J, Cladera G (1995) Application of computerized tomography for studying insect traces. Ichnos 4:77–81

Genise J, Farina J (2012) Ants and xenarthrans involved in a quaternary food web from Argentina as reflected by their fossil nest and palaeocaves. Lethaia 45(3):411–422

Genise J, Hazeldine P (1995) A new insect trace fossil in Jurassic wood from Patagonia, Argentina. Ichnos 4:1–5

Genise J, Hazeldine P (1998) 3 D reconstruction of insect trace fossils: Ellipsoideichnus meyeri Roselli. Ichnos 5:167–175

Genise J, Pazos P, González M, Tófalo R, Verde M (1998) Hallazgo de termiteros y tubos meniscados en la Formación Asencio (Cretácico Superior—Terciario Inferior) R.O. del Uruguay. III Reunión Argentina de Icnología y I Reunión de Icnología del Mercosur, Resúmenes: 12–13. Mar del Plata, Argentina

Genise J, Mángano M, Buatois L, Laza J, Verde M (2000) Insect trace fosil associations in paleosols: the Coprinisphaera Ichnofacies. Palaios 15:49–64

Genise J, Bellosi E, González M (2004) An approach to the description and intepretation of ichnofabrics in paleosols. In: McIlrroy (editor), "The application on ichnology to palaeoenvironmental and stratigraphic analysis" vol 228, Geological Society, London. Special Publications, pp 355–382

Genise J, Melchor R, Bellosi E, González M, Krause M (2007) New insect pupation chambers (Pupichnia) from the upper cretaceous of patagonia, argentina. Cretac Res 28:545–559

Genise J, Melchor R, Bellosi E, Verde M (2008a) Invertebrate and vertebrate trace fossils from continental carbonates. Dev Sedimentol 6:321–371

Genise J, Bedatou E, Melchor R (2008b) Terrestrial crustacean breeding trace fossils from the Cretaceous of Patagonia (Argentina): paleobiological and evolutionary significance. Palaeogeogr Palaeoclimatol Palaeoecol 264:128–139

Genise J, Alonso-Zarza A, Krause M, Sánchez V, Sarzetti L, Farina J, González M, Kosarinsky M, Bellosi E (2010) Rhizolith balls from the lower Cretaceous of Patagonia (Argentina): just roots or the oldest evidence of insect agriculture. Palaeogeogr Palaeoclimatol Palaeoecol 287:128–142

Genise J, Melchor R, Sánchez V, González M (2013) Attaichnus kuenzelii revisited: a Miocene record of fungus growing ants from Argentina. Palaeogeogr Palaeoclimatol Palaeoecol 386:349–363

Gonçalves C (1961) O Género Acromyrmex no Brasil (Hymenoptera, Formicidae). Estudia Entomol 4(1–4):113–180

González M (1979) Las líneas de costa de la Salina del Bebedero, Provincia de San Luis y su posible implicancia paleoclimática. Unpublished preliminary report, IANIGLA, Mendoza. (cited by Pascual, R and Bondesio, P, 1981)

González M, Tófalo O, Pazos P (1998) Icnología y paleosuelos del Miembro del Palacio de la Formación Asencio (Cretácico Superior-Terciario Inferior) del Uruguay. II Congreso Uruguayo de Geología, Actas, pp 38–42

Grassé P (1984) Termitología. Tomo II, Massón, París

Guiñazú J (1960) Los llamados "Estratos de Los Llanos" en la provincia de San Luis y su contenido de rocas andesíticas y restos de mamíferos fósiles. Primeras Jorn Geológicas Argentinas. Anales II: 89–95. Buenos Aires

Halffter G, Edmonds W (1982) The nesting behavior of dung beetles (Scarabaeinae): An ecological and evolutive approach. Instituto de Ecología, México D.F, p 176

Häntzschel W (1962) Trace fossils and problemática. In: Moore RC (ed) Treatise on invertebrate paleontology Pt. W, Miscellanea. Laurence, Kansas. Geological Society of America and University Kansas Press, pp W177–W245

Häntzschel W (1965) Vestigia invertebratorum et Problemática. Fossilium Catalogus, Animalia, pars 108

Hasiotis S, Bown M (1992) Invertebrate trace fossils: the backbone of continental ichnology. In: Maples C, West R (eds) Trace fossils short courses in Paleontology. N° 5, Paleontological Society. pp 64–104

Hasiotis S, Demko M (1996) Terrestrial and freshwater trace fossils, Upper Jurassic Morrison Formation, Colorado Plateau. In: Morales M (ed) The continental Jurassic, Museum of Northern Arizona, vol 60. Bulletin, pp 355–370

Hasiotis S, Dubiel R (1995) Termite (Insecta: Isoptera) nest ichnofossils from the Upper Triassic Chinle Formation, Petrified Forest Nacional Park, Arizona. Ichnos 4:119–130

Hasiotis S, Aslan A, Bown T (1993) Origin, architecture and paleoecology of the Early Eocene continental ichnofossil Scaphichnium hamatum, integration of ichnology and paleopedology. Ichnos 3:1–9

Hunt A, Lucas S (2007) Cenozoic teropod ichnofaunas and ichnofacies. New Mex Mus Nat Hist Sci Bull 44

Hunt A, Lucas S, Lockley G, Haubold H, Braddy S (1994) Tetrapod ichnofacies in early permian red beds of the American Southwest. New Mex Mus Nat Hist Sci Bull 6

Imbellone P, Teruggi M (1988) Sedimentación crotovínica en secuencias cuaternarias bonaerenses. Soc Argent de Sedimentología, Actas 2:125–129

Imbellone P, Teruggi M, Mormoneo L (1990) Crotovinas en sedimentos cuaternarios del Partido de La Plata. In: Zárate M (ed) Simposio Internacional sobre loess. Características, cronología y significado paleoclimático del loess. Actas. Mar del Plata, pp 166–172

Iriondo M, Krohling D (1996) Los sedimentos eólicos del noreste de la llanura pampeana (Cuaternario superior). XIII Congreso Geológico Argentino y III Congreso de exploración de hidrocarburos. Actas 4:27–48

Iriondo M, Krohling D (2001) A neoformed kaolinitic mineral in the Upper Pleistocene of NE Argentina. In: International clay conference. Abstract, 12: 6

Janet C (1898) Réaction alkaline des chambers et galleries des nids de Fourmis. Comptes Rendus Acad. Sciences. Paris, France, pp 130–133

Keighley D, Pickerill R (1994) The ichnogenus beaconites and its distinction from ancorichnus and Taenidium. Palaeontology 37:305–337

Klappa C (1980) Rhizoliths in terrestrial carbonates: classification, recognition, genesis and significance. Sedimentology 27:613–629

Kraus M, Hasiotis S (2006) Significance of different modes of rhizolith preservation to interpreting paleoenvironmental and paleohydrologica settings: examples from paleogene paleosols, Bighorn Basin, Wyomings, U.S.A. J Sediment Res 76:633–646

Krell R (2000) The fossil record of mesozoic and tertiary scarabaeoidea (Coleoptera: Polyphaga). Invertebr Taxonomy 14:871–905

Kusnezov N (1963) Zoogeografía de las hormigas en Sudamérica. Acta Zoológica Lilloana 19:25–186

Kusnezov N (1978) Hormigas argentinas. Clave para su identificación. Ilustraciones: I – XXVIII. Miscelanea, 61, Fundación Miguel Lillo, edited by Golbach R

Laza J (1982) Signos de actividad atribuibles a Atta (Myrmicidae) en el Mioceno de la Provincia de La Pampa, República Argentina. Significación paleozoogeográfica. Ameghiniana 19:109–124

Laza J (1986a) Icnofósiles de paleosuelos del Cenozoico mamalífero de Argentina. I. Paleógeno. Boletín Asociación Paleontológica Argentina 15:19

Laza J (1986b) Icnofósiles de paleosuelos del Cenozoico mamalífero de Argentina. II. Neógeno. Bol Asoc Paleontológica Argentina 15:13

Laza J (1995) Signos de actividad de insectos. In: Alberdi M, Leone G, Tonni E (ed) Evolución biológica y climática de la región pampeana durante los últimos cinco millones de años. Museo Nacional de Ciencias Naturales, Madrid. Monografía Nº 12, pp 347–361

Laza J (1997) Signos de actividad atribuibles a dos especies de Acromyrmex (Myrmicinae, Formicidae, Hymenoptera) del Pleistoceno en la provincia de Buenos Aires, República Argentina, significado paleoambiental. Geociencias II 6:56–62

Laza J (1998) Presencia de nidos de escrabajos en depósitos de la Cueva Tixi, partido de General Alvarado, Argentina. Significación paleoclimática y cronológica. Tercera Reunión Argentina de Icnología y Primera Reunión de Icnología del Mercosur. Mar del Plata. Resúmenes: 15

Laza J (2001) Nidos de Scarabaeinae. Significación paleoclimática y cronológica (pp 119–122). In: Mazzanti D, Quintana C (ed) Cueva Tixi: Cazadores y recolectores de las Sierras de Tandilia Oriental. 1. Geología, Paleontología y Zooarqueología. Universidad Nacional de Mar del Plata. Laboratorio de Arqueología. Publicación Especial 1

Laza J (2006a) Termiteros del Plioceno y Pleistoceno de la provincia de Buenos Aires, República Argentina. Significación Paleoambiental y Paleozoológica. Ameghiniana 43(4):641–648

Laza J (2006b) Dung-Beetle fosil brood balls: the ichnogenera coprinisphaera sauer and quirogaichnus (Coprinisphaeridae). Ichnos 13:1–19

Laza J, Tonni E (2004) Possible trace fossils of termites (Insecta, Isoptera) in the Late Cenozoic of the eastern Pampean región, Argentina. In: First international congress on ichnology. Abstract Book: 46. Trelew, Argentina

Laza J, Genise J, BownT (1994) Arquitectura y origen de Monesichnus Ameghino Roselli, revelada por tomografía computada. Reunión de Comun Asoc Paleont Arg Ameghiniana 31(4):397

Lockley M (2007) A tale of two ichnologies: The different goals and missions of vertebrate and invertebrate ichnology and how they relate ichnofacies analysis. Ichnos 13:39–57

Lockley M, Hunt P, Meyer C (1994) Vertebrate tracks and the ichnofacies concept: implications for paleoecology and palichnostratigraphy (241-268). In: Donovan S (ed) The Paleobiology of trace fossils. Wiley and Sons, New York

Lovelock JE (1979) Gaia, a new look at life on Earth. Oxford University Press

Manera T, Aramayo S (2004) Taphonomic features of Pehuen-Có palaeoichnological site (Late Pleistocene), Buenos Aires province, Argentina. First International Congress on Ichnology. Abstract book: 49. Trelew, Argentina

Manera T, Aramayo S, Ortiz H (2005) Trazas de pelaje en icnitas de megaterios en el yacimiento paleicnológico de Pehuen-Có (Pleistoceno tardío) provincia de Buenos Aires, Argentina. Ameghiniana, 42(4) Suplemento: 73R

Manera T, Bastianelli N, Aramayo S (2010) Nuevo registro de icnitas de mamíferos pleistocenos en Playa del Barco, Pehuen-Có, Provincia de Buenos Aires, Argentina. X Congreso Argentino de Paleontología & VII Congreso Latinoamericano de Paleontología, La Plata, Argentina. Universidad Nacional de La Playa, Museo de La Plata, Resúmenes: Nº 85: 78

Mángano M, Buatois L (2001) El Programa de Investigación Seilacheriano: la Icnología desde la perspectiva de Imre Lakatos. Asoc Paleontológica Argentina, Publicación Especial 8:177–186

Martínez S, Ubilla M (2004) El Cuaternario en Uruguay. In: Bossi L (ed) Geología de uruguay. Montevideo, pp 195–227

Martin L, West D (1995) The recognition and use of dermistid (Insecta, Coleoptera) pupation chamber in paleoecology. Palaeogeogr Palaeoclimatol Palaeoecol 113:303–310

Mazzanti D, Quintana C (eds) (2001) Cueva Tixi: Cazadores y recolectores de las Sierras de Tandilia Oriental. 1. Geología, Paleontología y Zooarqueología. Laboratorio de Arqueología. Universidad Nacional de Mar del Plata. Publicación Especial 1

Melchor R, Genise J, Miquel S (2002) Ichnology, sedimentology and paleontology of Eocene calcareous paleosols from a palustrine sequence, Argentina. Palaios 17:16–35

Melchor R, Bellosi E, Genise J (2003) Invertebrate and vertebrate trace fossils from a Triassic lacustrine delta: the Los Rastros Formation, Ischigualasto Provincial Park, San Juan, Argentina. In: Buatois, Mángano (ed) Icnología: Hacia una convergencia entre Geología y Biología. Asociación Paleontológica Argentina. Publicación Especial N° 9, pp 17–33

Melchor R, Bedatou E, de Balais S, Genise J (2006) Lithofacies distribution of invertebrate and vertebrate trace fossil assemblages in an Early Mesozoic ephemeral fluviolacustrine system from Argentina: Implications for the Scoyenia ichnofacies. Palaeogeogr Palaeoclimatol Palaeoecol 239:253–285

Melchor R, Genise, J, Buatois, L, Umazano, A (2012). Fluvial environments. In: Knaust, Bromley (eds) Trace fossils as indicators of sedimentary environments. Developments in Sedimentology, 64: 329–378

Mikulás R, Genise J (2003) Traces within traces: holes, pits and galleries in walls and fillings of insect trace fossils in paleosols. Geol Acta 1(4):339–348

Montalvo C (2004) Late Miocene coprolites from the Cerro Azul Formation at Caleufú, La Pampa, Argentina. First International Congres on Ichnology. Abstract Book: 58. Trelew, Argentina

Morrone J (2000) What is the Chacoan subregión? Neotropica 46:51–68

Morrone J (2001) Biogeografía de América Latina y el Caribe. Sociedad Entomológica Aragonesa, Zaragoza. Manuales y Tesis, vol. 3, p 144

Mouzo F, Farinati E. Espósito G (1985) Tubos fósiles de Callianassidos en la playa de Pehuen-Có, provincia de Buenos Aires. I Jornadas Geológicas Bonaerenses. Resúmenes: 87. Tandil

Nabel P, Machado G, Luna A (1990) Criterios diagnósticos en la estratigrafía de los "sedimentos pampeanos" del noreste de la provincia de Buenos Aires, Argentina. XI Congreso Geológico Argentino, Actas II, pp 121–124

Nabel P, Machado G, Luna A (1990) Criterios diagnósticos en la estratigrafía de los "sedimentos pampeanos" del noreste de la provincia de Buenos Aires, Argentina. XI Congreso Geol Argentino, Actas II: 121–124

Nathorst A (1881) Mémoire sur quelques traces d'animaux sans vertebres etc. et de leer portée paléontologique. Kgl. Svenka Vetensk. Akad. ANLD 18, 104 pp. (In Swedish, cited by Osgood, 1975)

Nesbit E, Campbell K (2005) The paleoenvironmental significance of Psilonichnus. Palaios 21(2):187–196

Noriega J, Areta J (2005) First record of Sarcoramphus Dumeril 1806 (Ciconiformes: Vulturidae) from the Pleistocene of Buenos Aires province, Argentina. J S Am Earth Sci 20:73–79

Osgood R (1975) The history of invertebrate ichnology. In: Frey RW (ed) The study of trace fossils. pp 3–12

Osterrieth M, Tassara D, Luppi T (2004) Estructuras biogénicas y restos fósiles en antiguos cangrejales en secuencias sedimentarias de la costa del Sudeste Bonaerense. X Reunión Argentina de Sedimentología. Resúmenes. San Luis, pp 121–122

Pagliarelli C, Bergqvist L, Maciel L (1993a) Icnofosseis de mamiferos na planicie costeira do rio Grande Do Sul. X Jornadas Argentinas de Paleontología de Vertebrados. Comun Resúmenes, Ameghiniana 30(3):323

Pagliarelli C, Bergqvist L, Maciel L (1993b) Icnofosseis de mamiferos (crotovinas) na Planicie costeira do Río Grande do Sul, Brasil. Anais Acad Bras Cienc 66(2):189–197

Paik I (2000) Bone chip-filled burrows associate with bored dinosaur bone in floodplain paleosols of the Cretaceous Hasandon Formation, Korea. Palaeogeogr Palaeoclimatol Palaeoecol 157:213–225

Plá S, Yebenes A, Soria J, Viseras C (2007) Carbonatos palustres en llanuras de inundación fluvial del Plioceno-Pleistoceno (Cuenca de Guadix, Granada, España). Geogaceta 43:107–110

Politis G (1993) Las pisadas humanas de Monte Hermoso dentro del contexto de la arqueología pampeana. Primera Reunión Argentina de Icnología. La Pampa, Argentina. Resúmenes y conferencias: 25

Pommi L, Tonni E (2010) Marcas de insectos sobre huesos del Pleistoceno tardío de la Argentina. X Congreso Argentino de Paleontología y Bioestratigrafía and VII Congreso Latinoamericano de Paleontología, La Plata. Universidad Nacional de La Plata, Museo de La Plata. Resúmenes (N° 342):210

Quintana C (1992) Estructura interna de una paleocueva, posiblemente de un Dasypodidae (Mammalia, Edentata) del Pleistoceno de Mar del Plata (provincia de Buenos Aires, Argentina). Ameghiniana 29(1):87–92

Quintana C, Martínez G, Osterrieth M, Mazzanti D (1998) Icnitas de mamíferos en un reparo rocoso de Tandilia Oriental (Pleistoceno tardío- Holoceno temprano), provincia de Buenos Aires. Tercera reunión Argentina de Icnología y Primera Reunión de Icnología del Mercosur. Mar del Plata. Resúmenes: 26

Ratcliffe B, Fagerstrom J (1980) Invertebrate lebensspuren of Holocene floodplains: their morphology, origin, and paleoecological significance. J Paleontol 54:614–630

Retallack G (1984) Trace fossils of borrowing beetles and bees in an Oligocene paleosol, Badlands National Park, South Dakota. J Paleontol 58:571–592

Retallack G (1990) Soils of the past. Unwin Hyman, Boston, p 520

Richter R (1927) Die fossilien fährten und Bauten der Wúrmer, ein überblick úber ihre biologischen Grundformen und deren geologische Bedeutung. Palaontologische Zeischrift, 9:193–240. (Mentioned in Osgood, 1975)

Rivas S (1900) Nueva teoría acerca de la formación geológica de algunas grutas del Uruguay. In: Araujo I (ed) Diccionario Geográfico del Uruguay. Imprenta Artística, Montevideo, p 548

Roberts E, Rogers R, Foreman B (2007) Continental insect borings in dinosaur bone: examples from the Late Cretaceous of Madagascar and Utah. J Paleontol 81:201–208

Roger R (1992) Nonmarine borings in dinosaur bones from the upper cretaceous two medicine formation, northwestern Montana. J Vertebr Paleontol 12(4):528–531

Roig V, Cox G (1985/86) La presencia de montículos tipo Mima en la Argentina en relación con roedores del género Ctenomys. Ecosur 12–13:93–100. Mendoza, Argentina

Roig V, González Loyarte M, Rossi M (1988) Ecological analisis of a mound formation of the Mima type in Río Quinto (province of Córdoba, Argentina). Stud Neotropical Fauna Environ 23:103–115

Roselli F (1939) Apuntes de geología y paleontología uruguaya. Sobre insectos del Cretácico del Uruguay o descubrimientos de admirables instintos constructivos de esa época. Bol de la Soc Amigos de las Cienc Nat "Kraglievich-Fontana" 1:72–102

Rossi V, Osterrieth M, Martínez G (2001) Icnología de la secuencia estratigráfica Mar Chiquita (Cuaternario tardío), Buenos Aires. IV Reunión Argentina de Icnología y Segunda Reunión de Icnología del Mercosur. Resúmenes: 68. Tucumán

Rusconi C (1937e) Contribución al conocimiento de la geología de la ciudad de Buenos Aires y sus alrededores y referencias de su fauna. Actas Acad Nac de Córdoba X:117–133

Rusconi C (1967) Animales extinguidos de Mendoza y de la Argentina. Edición Oficial, 46 lám. Mendoza, pp 489

Sánchez V (2009) Trazas fósiles de coleópteros coprófagos del Cenozoico de la Patagonia Central. Significado evolutivo y Paleoambiental. Unpublished doctoral tesis, Facultad de Ciencias Exactas y Naturales, Universidad Nacional de Buenos Aires

Sánchez V, Genise J (2009) Cleptoparasitismo and detritivory in dung beetle fossil brood balls from Patagonia, Argentina. Palaeontology 52(4):837–848

Santa Cruz M (1979) Geología de las unidades sedimentarias aflorantes en el área de las cuencas de los ríos Quinto y Conlara, provincia de San Luis, República Argentina. VII Congreso Geológico Argentino, Actas I: Neuquén, pp 335–349

Sauer W (1955) Coprinisphaera ecuadoriensis, un fósil singular del Pleistoceno. Bol del Inst de Cienc Nat Ecuador 1:123–132

Schutze J (1907) Die Lagerungsverhalttenisse Bunter Breccie an der Bahlinie Donaukworth-Treuchtlingen und ihre Bedeutung fur das riesproblem. In: Branca W, Fraas E (ed) Berlin, 2. Cited in Genise, 1999, pp 25–26

Scognamillo D, Zárate M, Busch C (1998) Estructura de las cuevas de Actenomys (Rodentia: Octodontidae) del Plioceno tardío (Barranca de Los Lobos, Partido de General Pueyrredón): significado paleoecológico y estratigráfico. Tercera Reunión Argentina de Icnología & Primera Reunión de Icnología del Mercosur. Mar del Plata. Resúmenes: 28

Seilacher A (1953) Studien zur Palichnologie. I. Über die Methoden der Palichnologie. Neues Jahrbuch für Geologie und Paläontologie, Abhandlungen 96(3):421–452

Seilacher A (1964) Biogenic sedimentary structures. In: Imbrie E, Newell S (eds) Approaches to paleoecology. Wiley, New York, U.S.A, pp 293–316

Seilacher A (1967) Bathymetry of trace fossils. Mar Geol 5:413–428

Silvestri F (1903) Contribuzione alla conoscenza dei Termiti e Termitofili dell America Meridionale. REDIA 1:1–234

Smith R, Mason T, Ward J (1993) Flash-flood sediments and ichnofacies of the Late Pleistocene Homeb Silts, Kuiseb River, Namibia. Sed Geol 85:579–599

Snyder T (1949) Catalog of the termites (Isoptera) of the world. Smithsonian Miscelanea Colloque 112(3953):1–490

Subachev V (1902) On the problem of crotovinas. Pochvovedenie 4:397–423

Tassara D, Aramayo S, Osterrieth M, Scian R (2005) Paleoicnología de mamíferos de la Formación Santa Clara (Pleistoceno Tardío) en la zona costera del Partido de Mar Chiquita (Provincia de Buenos Aires), Argentina. XVI Congreso Geol Argentino, Actas, pp 257–258

Terry R (2004) Owl pellet taphonomy: a preliminary study of the post-regurgitation taphonomic history of pellets in a temperate forest. Palaios 19(5):497–506

Teruggi M, Bianchini J, Tonni E (1972) Un cuerpo crecionario de origen animal que semeja un huevo fósil. Rev Asoc Geol Argentina 27(4):391–398

Todd J (1903) Concretions and their geological effects. Bull Geol Soc Am 14:353–368

Tonni E, Verzi D, Bargo M, Pardiñas J (1993) Micromammals in owl pellets from the Lower-Middle Pleistocene in Buenos Aires Province, Argentina. X Jorn Argentinas de Paleontología de Vertebrados. Resúmenes. Ameghiniana 30(3):342

Trackray G (1994) Fosil nest of sweat bees (Halictinae) from a Miocene paleosol, Rusinga Island, Western Kenya. J Paleontol 68(4):795–800

Vaz Ferreira R, De Zolessi C, Achaval F (1970) Oviposición y desarrollo de ofidios y lacertilios en hormigueros de Acromyrmex. Physis 29(79):431–459

Verde M, Ubilla M (2002) Mammalian carnivore coprolites from the Sopas Formation (Upper Pleistocene, Lujanian Stage), Uruguay. Ichnos 9:77–80

Verde M, Ubilla M, Jiménez J, Genise J (2006) A new earthworm trace fossil from paleosols: aestivation chambers from the Late Pleistocene Sopas Formation of Uruguay. Palaeogeogr Palaeoclimatol Palaeoecol 243:339–347

Vizcaíno S, Zárate M, Bargo M, Dondas A (2001) Pleistocene burrows in the Mar del Plata area (Argentina) and their probable builders. Acta Paleontol Polonica 46(2):289–301

Voglino D (1999) Geología superficial y paleontología de las barrancas del Río Paraná entre Rosario (Santa Fé) y Campana (Buenos Aires). Private edition, Buenos Aires

Weber N (1972) Gardening ants: the Attines. The American Philosophical Society, vol 92. Philadelphia, U.S.A, pp 146

Wheeler W (1910) Ants; their structure, development and behavior. Columbia University Press

Zárate M, Bargo M, Vizcaino S, Dondas A, Scaglia O (1998) Estructuras biogénicas en el Cenozoico tardío de Mar del Plata (Argentina) atribuibles a grandes mamíferos. Asoc de Sedimentología 5(2):95–103

Zárate M, Schultz P, Blasi A, Heil C, King J, Hames W (2007) Geology and geochronology of type Chasicoan (late Miocene) mammal-bearing deposits of Buenos Aires (Argentina). J S Am Earth Sci 23:81–90

Zavala C, Navarro E (1993) Depósitos fluviales en la Formación Monte Hermoso (Plioceno Medio-Superior), provincia de Buenos Aires. XII Congreso Geol Argentino II de Exploración de Hidrocarburos, Actas 2:236–244

Zavala C, Quattrocchio M (2001) Estratigrafía y evolución geológica del río Sauce Grande (Cuaternario), provincia de Buenos Aires, Argentina. Rev de la Asoc Geol Argentina 56(1):25–37

Zorn M, Genise J, Gingras M, Pemberton S (2010) Wall types for trace fossils: the micromorphological perspective. Unpublished report

Chapter 3
Faunistic Associations and Climatic Events During the Pan-Araucanian Cycle

Abstract The variations that the diastrophic events, as well as the climatic changes, inflicted to the Pampean Province gave special geographic and biotic features. The study of these historical events was divided by specialists, especially vertebrate pale-ontologists, in two big cycles, the Pan Araucanian and the Pan Pampean. Each of these vast time spans were subdivided in stages, which carry fossil elements of determined and specific taxa. In addition to the fruitful paleovertebrate work, there was research done by paleobotanists and detailed work by geologists and sedimentologists. The latter then added the study on paleosols. Finally, the ichnological studies began to provide certain details about the biota which circulated and inhabited those pale-osols. This new set of information verifies numerous statements provided by previous researchers and, in turn, it allows specifying some subtle biological–climatic details.

Keywords Cycles · Stages · Paleosols · Biota · Climate

Pascual and Bondesio (1982) pointed out that the diastrophic phases of the late Miocene influenced the origin of a huge plain system, substituting the retreating "Paranaense Sea." Such plains spanned from the current region of the Valdés Penin-sula to the Gran Pantanal region in the Brazilian Matto Grosso, establishing the "Age of Southern Plateaus" during the Chasicoan to the Marplatan stages. Thus, during this long time, a series of environments similar to those corresponding to the biotic provinces of the Chaqueña Sub-region of Morrone (2000, 2001) took place in Pampasia.

Except for the outcrops corresponding to the Huayquerian stage in the Argentine Mesopotamia area (provinces of Misiones, Corrientes and Entre Ríos, eastern Argentina), the common characteristic of these units is their sedimentological homo-geneity. These sediments are fine sands and silts, that appear forming layers, mostly edaphic, developing true successions of paleosols, some of which carry different carbonatic formations and conspicuous levels with "escorias" (slags) and "tierras cocidas" (earthen cooked).

The loessic plains result from the climate and sedimentation (Teruggi and Imbel-loni 1988). The profiles of overlapped paleosols represent thousands of years of sedimentation and pedogenesis, making them exponents of the past climatic actions.

© The Author(s), under exclusive license to Springer Nature Switzerland AG 2020   69
J. H. Laza, *Ichnology of the Lowlands of South America*, Springer Earth System Sciences,
https://doi.org/10.1007/978-3-030-62597-9_3

The hydrologic changes are those best appreciated, since they translate into soil morphology and micromorphology. Besides, the presence of different associations of fossil vertebrate remains allowed the elaboration of a reasonable chronologic sequence and finally, the growing data availability from isotopic chronology, which provides better accuracy to temporal lucubration.

The Andean orogeny process of the Tertiary age is one of the most significant and deepest vicariance events in South America, shown by the biota fragmentation—originally very widely distributed—which forms sets of trans-Andean and regional sibling groups, especially at high taxonomic levels, as one of the causes of South American plant and zoological richness (Croizat 1976). This fragmentation is multiplied by the development of great hydrological networks (Haffer 1969). Thus, savanna environments follow in Pampasia, alternating with grassland ecosystems. Beginning in the Cenozoic, the grassland ecosystem undergoes, seasonally, the action of natural fires, where herbs with their underground rhizomes quickly recover their development (Retallack 2001). Eiten (1997), when typifying savannas, recognized that there are two kinds: (a) the physiognomic vegetation form and (b) where such term is used for characterizing the formation where flora, climate and substratum take part. Both groups include savannas which undergo periodical natural fires. The savannas in South America receive enough annual rainfall (more than 1000 mm/yr) distributed in more than seven months, for sustaining ever green mesophilic greenwoods. Eiten (1972), when referring to the action of fire on the vegetation, stated that they have interacted for a long time, forming a group of well-adapted ecological types. Sarmiento (1992), when classifying savannah ecology in reference to those in South America, distinguished that in semi-arid tropical climates, tree-covered forests or formations of thorny bushes develop not always together with a continuous litter of perennial graminoids.

Morrone (2000–2001), based on the trace integration and the reconnaissance of ecoregions, proposed a renovating diagram of the Neotropical region biogeography, assigning the degree of the Chaqueña Sub-region to the previous Chaqueño Domain of Cabrera and Willink (1973). Such sub-region is located within the subtropical belt, with annual average temperatures ranging from 10 to 24 °C and decreasing from north to south. In summer, the rain regime, seasonal, concentrates 80% of water, recording rainfall ranging from 1200 mm in the east to 450 mm in the southwest, and that due to an orographic effect they increase over Pampean and Sub-Andean mountains that are in the western border (Bucher 1980).

This region is formed by the following provinces: (1) **Caatinga,** whose vegetation comprises different kinds of xeric environments, from dry woods to open scrubland and savannahs with cactus, developing wetter woods in elevations higher than 500 m; (2) **Cerrado**, one of the largest savannas/woodland complexes in the world. It has open woods with trees up to 12 m tall, a bush stratum and another one of herbs, rich in gramineae and legumes. Redford (1986) stated that 50% of the mammal fauna in Cerrado comes from forests and savannahs, due to woods in gallery that act as corridors. (3) **Chaco**, populated by deciduous, xeric woodlands, with a stratum of gramineae, cacti and terrestrial bromeliads; savannas and halophilus steppes are also frequent. (4) **Monte**, populated by open scrubs and cacti with gramineae patches, and (5) **La Pampa,** which be described when dealing with the Pan-Pampean cycle.

The main adaptive problems of the Chaqueña fauna, like in all semi-arid regions, are mainly associated to maintaining the thermal and hydrological balance against extreme conditions, such as the defense of fires and flooding and to the exploitation of resources that appear seasonally and irregularly. Thus, the caves constitute a microhabitat that offers excellent protection against high temperatures and desiccation, and the loose soils of Chaco do not represent a hindrance for excavation. The cavicola habits are very spread in the Chaqueña fauna, including a great number of medium and small vertebrates, which share the living space with invertebrates in some cases. Abandoned caves are also occupied.

The paleofloristic information, in addition to different associations of fossil vertebrates, allowed inferring environmental conditions for each of the recognized stages and sub-stages, information that will be contrasted with ichnological findings as it follows.

3.1 Chasicoan (Fig. 3.1)

The area under study shows scarce sites of this age, one of them (locus tipus), circumscribed to the depression of Laguna (i.e., shallow lake) Chasicó and the valley of Arroyo (i.e., creek) Chasicó in its mouth. Lithological and mainly subsequent paleontological studies subdivided the formation originally named as the Chasicó Formation (Pascual 1961) in two members: Vivero (lower) and Las Barrancas (upper) members (Bondesio et al. 1980; Fidalgo et al. 1987). From the paleontological point of view, two biozones are recognized: (a) Chasicotherium rothi (Tonni et al. 1998) and (b) Chasicotatus ameghinoi (Cione et al. 2000); a third biozone, Chasichymys bonaerense, was added by Verzi et al. (2008).

Sedimentation represents the development of fluvial environments of channel bars and plain areas of flooding and overflow with intervals of soil formation with root casts, carrier of glassy and vesicular slabs ("escorias") and field burnings ("tierras cocidas"). Subsequent studies in the stratotype (Zárate et al. 2007) mentioned the presence of paleosols with root casts, invertebrate bioturbations and some excavations of 30 cm in diameter, without specifying more about these topics.

The study of the Chasicoan stage faunal group led to infer open spaces with scarce tree vegetation and grasslands subjected to seasonal variations (Bondesio et al. 1980).

3.2 Huayquerian (Fig. 3.2)

The Huayquerian stage, extensively represented in the region, developed in both watersheds of the "Paranaense" Sea. The type area of the early Huayquerian stage is in the lower course of the Arroyo Chasicó, in Buenos Aires Province, divided

Fig. 3.1 Chasicoan—Chasicó shallow lake

in biozones of Macrochorobates scalabrinii (Tonni et al. 1998) and Chasichymys scagliai (Verzi et al. 2008). Later studies about the evolutionary state of rodents found in coeval sediments in La Pampa Province encouraged the creation of four new biozones: morphotype Chasichymys, Xenodontomys simpsoni, Xenodontomys ellipticus and Xenodontomys elongatus (Verzi et al. 2008). During the Huayquerian stage, the "pan-Araucarian" faunal cycle started. Its sediments extended from the Valdés Peninsula in Chubut toward the north through numerous piedmont and inter-mountainous sites that extend, at least, as far as the Andean foothills of the Bolivian plains. In the study area, its appearances are placed in two watersheds of the "Paranaense" Sea: (a) the gullies surrounding the Río Paraná in its medium and lower course, in the Corrientes and Entre Ríos provinces, and north of Uruguay, and (b) La Pampa, San Luis, Córdoba and west of Buenos Aires provinces. Both regions develop different biotic paleocommunities.

a. Río Paraná. Sediments of the Ituzaingó Formation ("Mesopotamiense") develop floodable plains, rain forest in galleries and fluvial deposits resulting from the "proto-Paraná" and "proto-Uruguay" activity. In the Entre Ríos Province, the layers deposited in paleobeds of the Paraná Formation receive the name of "conglomerado osífero" (i.e., bone bed conglomerate) for the content and high diversity of fossil remains that record, such as continental fish, a rich herpetofauna represented by turtles, saurians and snakes and among the Crocodilia, Caiman species, Eusuchia and Gavialidae (Gasparini and Báez 1975), birds and mammals. The Terror bird (Andalgalornis) populated open zones (Tonni 1980), while rodents hydroquerids, dinomids and neoepiblemids were at their peak.

Fig. 3.2 Huayquerian—Salinas Grandes de Hidalgo cliffs (photo by EP Tonni)

Their diversity is similar to the present tropical rain forests (Vucetich and Verzi 1999). Litopterna and Notoungulata represented by proteroterids, macrauquenids and toxodontids, only record grazing forms, while Xenarthra, represented by Megalonychops, are not found at the eastern slope.

The vegetation in this sector, separated from the Cordillera area by the "Entrerriense-Paranaense" Sea develops the strong influence of the Brazilian hydrophilic forests. The paleoxilological studies in the region represent the Fabaccac and Anacardeaceae families by imprints of leaves and fruits, logs, cyclusphaera and phytoliths (Barreda et al. 2007).

a. La Pampa Province and western Buenos Aires Province. The base of the sequence is represented by two formations:
b. The Río Negro Formation (Andreis 1965) composed of volcanoclastic materials deposited in alluvial and eolian environments in the basin of the Negro and Colorado rivers, reaching southern Buenos Aires Province.

  1. The Cerro Azul Formation (Linares et al. 1980) that appears in the Pampasia area in several localities of La Pampa, western Buenos Aires Province and occasionally in the foothills of the mountain ranges of the Sierras of San Luis and Córdoba. Its stratigraphic analysis in sites such as the Salinas Grandes de Hidalgo and Barrancas Coloradas in La Pampa Province (Montalvo et al. 1998, Goin et al. 2000) recognized three sedimentary phases as follows: (a) lake deposits that evolved from deep to shallow, carriers of desiccation track and pedogenetic levels of rich and varied fauna and fossil ichnofauna; (b)

eolian levels of greater distribution, with calcareous pellets and pedogenetic evidence, insect nests and fossils; (c) deposits of river stream beds. The pale-osols correspond to calcium vertisols (Melchor et al. 2000). As a consequence of the continental union through the Panama isthmus during the development of the Diaguita Diastrophic Phase, and the beginning of the Great American Biotic Interchange (Cione et al. 2015), the Cerro Azul Formation is a carrier of remains of the first immigrants coming from the Northern Hemisphere (Procyonidae, Mustelidae and Cricetidae rodents) (Verzi and Montalvo 2008; Cione et al. 2015).

During the Huayquerian stage, the herpetofauna in the La Pampa Province was represented by colubrids and vipers (Albino and Montalvo 2006) as well as by the lizard Tupinambis (Albino and Montalvo 2006), the latter recorded also in the province of Córdoba (Donadío 1984). At the same time, a varied avifauna populated environments of La Pampa and western Buenos Aires Province, sheltering open space inhabitants such as Theratornithidae and Phorusrhacidae and tinamids of wooded areas (Tonni 1980; Cenizo and Montalvo 2006).

Marsupials, during their last adaptive radiation in South America (Pascual and Bond 1986), were represented by Didelphidae, inhabitants of wooded areas, together with the first Thylacosmilidae (marsupial sabertoothed tigers) and Argyrolagidae, small rodentiform forms, inhabitants of grassland areas (Simpson 1970a).

Litopterna, represented by proteroterids, inhabitants of forested areas and macraucheniidae of grazing regime, as well as Notongulate represented by meso-terids, toxodontids, the largest hegetoterids and abundant and varied Pachyrukhinae (Bond 1999).

At that moment, Xenarthra reached the highest climax and diversity of their history (Carlini and Scillato-Yané 1999). Armadillos, as well as myrmecophagidae and purely arboreal cyclopids indicate tropical to subtropical conditions of the "Chaqueño" type (Scillato-Yané 1986). Carnivores were represented by procionids on both coasts of the "Paranaense" Sea. In turn, caviomorph rodents presented diver-sity similar to that of the Neotropical rainy forests (Montalvo et al. 1998); according to the finding of Eumysopinae in Barrancas Coloradas (La Pampa Province), at present restricted to the Chaqueña sub-region (Morrone 2000–2001), where there are rain forests as far as open areas of Cerrado and Caatinga.

The ichnological record of several sites of Huayquerian age is represented by two continental ichnofacies: (a) Scoyenia ichnofacies, represented in the following formations: (1) the Río Negro Formation, from which several vertebrate tracks were described (Megatheriinae, ungulates, marsupials and phororhacoid birds) in envi-ronments of interdune, temporal shallow lakes, with ripple marks and desiccation cracks; (2) the Cerro Azul Formation, from which the ichnogenus Taenidium was mentioned, as a result of the activity of warms that inhabit wetter soils and period-ically migrate to different surface levels according to their biological cycle; (3) the Ituzaingó Formation deposits, carriers of countless coprolites of fish, reptiles and terrestrial vertebrates preserved in environments from very wet to fully aqueous. Despite their number and variety, they were mentioned only recently.

Coprinisphaera ichnofacies presents abundant ichnodiversity, and it is represented in the Cerro Azul Formation by:

1. Coprolites whose preservation can only be guaranteed by a fast burial, in an aqueous medium or by eolian sedimentation.
2. Crotovines, in this case of large size, attributed to edentates. Their construction in open zones well may serve as shelter and progeny breeding.
3. Rhizoliths, frequent in the numerous and strong paleosols, in the shape of carbon pellets or ducts covered with manganese. Their abundance suggests a rich and extended vegetable cover.

The Cerro Azul Formation is abundant in the record of insect activity in paleosols, being represented by three ichnofamilies:

Celliformidae: cells built by bees, which require surfaces relatively clear of creeping vegetation for making their nests.

Krausichnidae: anthills attributable to Attini tribe were discovered in paleosols of the Cerro Azul Formation; one of them with remains of its inhabitants and a squamata guest.

1. It corresponds to anthill remains of the genus Atta (Attaichnus kuenzeli), builders of great living structures up to 2 million individuals. Herbivores prevailing in the Neotropic, the insects are the only animals that develop mutualism with fungi. They perform an intense foraging on grasslands and trees for developing their feeding strategy, where each species is monophagous as regards eating only one kind of fungi.
   The tribe distribution spans from 44° S latitude to 40° N latitude, on the eastern coast of North America, whereas Atta genus spans from 30° N latitude to 32° S latitude.
2. The other anthill, also attributed to the Attini tribe, corresponds to a reduced structure, assignable to a colony of Acromyrmex genus, whose nests are often shelter and protection for oviposition of several squamata, whose presence was detected inside the fossil structure together with inhabitant remains and progeny cocoons.

The reduction process of the litterfall mass of certain plants is practiced in the Neotropical Region by Attini ants, whose nesting activity produces alterations in the soils due to the material transport of soil deep horizons to the surface. The genera Trachymyrmex, Acromyrmex and Atta—the most important of the tribe—represent consecutive stages of a same lineage, while the other genera are lateral ramifications of an unknown common phylogenetic branch (Kusnezov 1963). Atta live in the wet tropical zone, with south boundary at 32° S latitude while great part of Acromyrmex, in extratropical zone, participate of the conquer process of xeric environments by originally mesophilic elements; historical migratory current in South America is in a northeast-southwest general direction (Kusnezov 1963). Some Acromyrmex forage on dicotyledonous, while others are specialized on gramineae. In certain environments, the woody vegetation only develops from Atta tumuliform anthills;

Iriondo (2007), when referring to the Santa Fe-Chaco geomorphology, stated that the northeast sector of the landscape is characterized by the presence of islets—resulting from the attachment of large anthills of the Atta genus—covered by tree and bush forests outstanding from the grassland landscape. Attini anthills are used by ophidian and lacertilian animals as protection and ovipositional sites, as described by Vaz Ferreira et al. (1970–1973) and Ferreira Brandao and Vanzolini (1985).

Coprinisphaeridae: specimens of two ichnogenera of coprophagus beetles: Coprinisphaera murguiai and Quirogaichnus coniunctus were found.

The adults of the subfamily Scarabaeinae accumulate provisions for the development of their offspring building chambers of specific size and characteristics that express the habit diversity. Even when most of Scarabaeinae are coprophagous specialized in excrement of large herbivores, there are saprophagous and necrophagous species, especially in the Neotropical region. According to the form of exploiting the food resource, there are three kinds of activity: (a) rolling or telecoprids, the adult couple sets aside a portion from the food pile, forms a ball that they roll at certain distance, lay an egg inside, cover the pill with a sediment layer and bury it; (b) diggers or paracoprids, the adult males transport the food into holes previously dug, where females build a ball, lay an egg and cover it with sediment; (c) endocoprids, the adults build directly their nest in the food pile. A large part of the species receives parental care and the complete development of individuals takes place within the chamber. The present study of these habits and their architectonic result allowed recognizing seven nesting patterns (Halffter and Edmonds 1982), from which the fossil record recognizes Patterns II, III, IV and V. Coprinisphaera murguiai corresponds to Pattern II and Quirogaichnus coniunctus corresponds to Pattern V. In the Neotropical region, diverse guilds of dunghill beetles associated to grasslands, forests and intermediate zones are distinguished. As part of a dominant insect group in the tropical regions, they inhabit lands with annual rainfall higher than 250 mm and average annual temperature higher than 15 °C.

3.3 Montehermosan (Fig. 3.3a, b)

Montehermosan stage sediments outcrop at the Buenos Aires Atlantic Ocean coast near the Sierra de la Ventana region and in streams and rivers draining that sector, appearing also in the provinces of La Pampa, San Luis and Córdoba. These deposits are related with periodically floodable plains, formed by successive paleosols, carriers of root tracks and the presence of "slag" and "terra cotta."

Zavala and Navarro (1993) pointed out that the sediments of both units that compose the Monte Hermoso cliffs (Montehermosan and lower Chapalmalalan stages) were deposited by the water dynamics of highly winding rivers, with high suspended load draining toward the west-southwest from the Sierras Bonaerenses positive craton. Discontinuity between both units is not significant and appears

(a)

(b)

Fig. 3.3 Montehermosan—Monte Hermoso marine cliffs **a** upper section and **b** lower and upper sections

several times in the outcrop, being related with deposits of flooding plains. Laminar deposits may be the answer to cyclic variations, maybe seasonal.

Among fossil vertebrates, the frogs Bufo paracmenis and Ceratophrys are present, the latter fossorial element of the Chaqueño xeric environments (Báez 1986) together with the lizard Callopistes bicuspidatos (Chani 1976), a genus whose more southern extant representatives reach the parallel 36° S and is not present in Argentina (Peters and Donoso Barros 1970). Also, the giant turtles Geochelone gallardoi are present, with ecological requirements of temperatures higher than 10 °C and the necessary shadow of wooded areas (Gasparini et al. 1986).

The bird record points out the last Teratornithidae appearance, at the same time that the presence of Rheidae, Cariamidae, Vulturidae and Tinamidae (Tonni and Tambussi 1986) has been mentioned.

Marsupials accounted for the last Caenolestidae, the persistence of Thylacosmilidae and the Argyrolaguidae.

Notoungulates appear reduced to three families: hegetotheridae, toxodontidae and the largest recorded mesotheridae.

The paquiruquids abounded together with xotodontids and Trigodon gaudryi, haploronterid with a frontal protuberance (Bond 1999). Litoptern macrauquenid inhabit wooded zones and grasslands (Bond 1999), whereas Xenarthra record large eufractins, vermilinguos and the first dasipodids which testify the persistence of the Chaqueño type conditions. The procionid Cyonasua, the Conepatus and the first felines show the greatest diversity among Carnivora.

Rodents point out the first miocastorids and ctenomids of great diversification, cavids to hidrocherids of cursorial formation and the first cricetids with grazing forms (Vucetich 1986). Besides, the first sigmodontines—from the Northern Hemisphere—are recorded in the biozone of Trigodon gaudri (Pardiñas 1999).

The ichnological examples for the Montehermosan stage are varied, corresponding to two ichnofacies: (a) Scoyenia ichnofacies. Smooth tubes and meniscal holes that Zavala and Navarro (1991) attributed to ichnogenus Muensteria and Taenidium made by warms or similar organisms at edaphic levels with high humidity content; vertebrate footprints similar to those attributed to megateroid edentates of bipedal locomotion (Casamiquela 1974) in the Río Negro Formation. (b) Coprinisphaera ichnofacies.

Crotovines, mentioned by Zavala and Navarro (1991) for the Montehermosan stage (Monte Hermoso cliffs). They described vertebrate caves measuring 18 cm to 1 m in diameter and 4 m long, with inclinations, ramifications and widening.

There are also two ichnofamilies in paleosols of the Monte Hermoso Formation.

1. Celliformidae, represented by Celliforma isolated cells.
2. Coprinisphaeridae, represented by the genus Coprinisphaera with this species, C. murguia, populating form of open environments, exploiter of coprophagia resources.

3.4   Chapadmalalan (Fig. 3.4)

The typical outcrop corresponds to the Farola de Monte Hermoso to the east of the city of Bahía Blanca, in Buenos Aires Province. The geological unit is formed by the upper member of the Monte Hermoso Formation of Zavala and Navarro (1991), part of the outcrops of the Irene Formation and elements of the Chapadmalal Formation of Kraglievich (1952). These sedimentary deposits are related to the plain formation and present much edaphic evidence. (Zárate et al. 1998).

The fish recorded in the Farola de Monte Hermoso site are Siluriformes Pimelodidae, members of the icthyofauna "Parano-Platense" (Ringuelet 1975) that together with percictids are the only continental fish known for the Pliocene in Argentina (Cione and Báez 2007). Besides, Gasparini and Báez (1975) mentioned the presence of Bufo pisanoi. The bird record makes us think about environments, though with trees, with bigger open spaces, whereas the best represented birds correspond to several species of tinamids; Tonni (1980) and Tambussi (1995) mentioned the record of different bird groups: reids, cariamids, catartids, charadrids and scolopacids.

Among marsupials, the carnivorous Thylatheridium and Sparassocynus were recorded and the tilacosmilids and the argirolaguids were also found.

Notungulates are represented by mesoterids, toxodontids, hegetoterids and numerous Paedotherium of probable cursorial-jumping habits (Bond 1999). Litopterna Brachytherium and Promacrauchenia appear as well. Litopterna show the last proterotherid record (maybe for the arrival of Holarctic ungulates), while

Fig. 3.4 Chapadmalalan—Chapadmalal marine cliffs (photo by EP Tonni)

Notongulates are represented by xotodontins and the probable appearance of the genus Toxodon (Bond 1999).

The record of Xenarthra species suggests less subtropical influence and development of semi-arid conditions with some tree vegetation (Carlini and Scillato-Yané 1999). Cingulate Xenarthra, represented by genera that exist nowadays (Chaetophractus, Zaedyus and Tolypeutes) together with pampaterids, consumers of abrasive vegetables (Edmond 1985) and varied glyptodontids. The present hairy tardigrada are notroterins and scelidoterins. The absence of myrmecophagidae and the comparison with different associations suggest the fall of average temperatures, as well as the prevalence of aridization conditions of the regional environments (Carlini and Scillato-Yané, 1999). Among carnivorous, the procionid Cyonasua is recorded for the last time.

Carnivorous appear diversified with the appearance of Holarctic elements, recording procionids (Chapalmalania) and the first macairodontino felids. Also, the first Tayassuidae appear coming from North America.

The environments of the Chapadmalalan stage show an increase in aridization and the prevalence of open areas, as evidenced by the impoverishment of rodent fauna (Vucetich and Verzi 1999).

During the Chapadmalan stage, a flora integrated by neotropical and southern elements was developed, characterized by herbaceous-bushy xerophytic vegetation (Barreda et al. 2007). In the Paraná River Basin, the deposits of the Ituzaingó Formation have paleobotanic and phytolithic elements demonstrating differences with the Miocene, though the same taxa are represented, showing the influence of the Paraná and Uruguay rivers (Barreda et al. 2007). Three types of vegetable communities are present: freshwater, stratified hygrophilous rain forest and xerophytic forest (Zucol et al. 2005). According to Aceñolaza and Aceñolaza (2004), some silicified logs present perforations, possibly produced by xilophagous organisms. In the La Pampa area, a flora integrated by Neotropical and southern elements developed, characterized by herbaceous-bushy xerophytic vegetation (Barreda et al. 2007). Linked to this information, it has been mentioned that at level VIII of the sequence described by Kraglievich (1952), it is carrier of an abundant and conspicuous amount, of calcareous pellets, sometimes mammillated and sometimes ramified, of great size. The presence of these elements reminds what Kraus (1988) expressed for the lower Tertiary of Wyoming, USA., where nodules at specific levels are abundant in paleosols of the Willwood Formation, forming bands that follow laterally channel sandstones without being inside the deposit itself (Bown and Kraus 1987). Some appear erect, cylindroid in shape of 1 m tall to 10–30 cm in diameter. In some cases, associated to tree trunks, they suggest that they are just badly preserved trunks.

It is also significant the presence of the sequence "slag" and "terra cotta" in several paleosols.

The ichnological stock of the Chapadmalan Stage appears represented by two continental ichnofacies:

a. Scoyenia Ichnofacies

 (1) Smooth tubes and meniscal holes of Taenidium, related to soil levels nearby the city of Bahía Blanca and the locality of Chapadmalal, Buenos Aires Province.

b. Coprinisphaera Ichnofacies

Several insect nests related with two subfamilies were recorded.

1. Coprinisphaeridae. Scarabaeinae (Coprinisphaera) nests are relatively frequent in paleosols of the Irene Formation and correspond to two ichnospecies: C. murguia and isp A (Sánchez 2009) whose specimens are relatively small.

The countless crotovines given at certain levels of paleosols have been recognized as one of the characteristics of this floor. The cave frequency of small mammals is evidence of an abundant rodent (caves with Actenomys of Genise 1989) and edentate faunas. The fossorial activity is important in the life of mammals, where great part of the families have at least a digging species, especially in environments with unfavorable climatic periods. In turn, they offer, through their living places, shared shelters or they occupy abandoned places, such as the mentioned case of edentate caves later inhabited by marsupials. It has also been detected inquilinism cases. In underground rodent burrows, small ball-nests of Scarabaeinae (Coprinisphaera) have been found, similar to those built at present by the canthonino Tetraechma sanguineomaculata, "a very common species in the Chaqueña zone" (Martínez 1959), whose present distribution is the northern Argentine provinces of Santiago del Estero, Chaco, Tucumán, Salta, Jujuy and Formosa, as well in Bolivia and Paraguay, living in "vizcacha" caves and using its excrement for nesting (Pereira and Martínez 1956).

The Krausichnidae ichnofamily is represented by ants (a, b, c) and termite nests (d).

2.3 Aff (Acromyrmex [Acromyrmex] ambiguus) ichnogenus. Attini ant genus inhabiting the Argentine Mesopotamia region, Río Grande do Sul (Brazil) and Uruguay. It settles in sandy hills of the Paraná–Uruguay river basins, building nests that have a light tumulus with a wide central mouth hidden by detritus. They have a major vaulted chamber, of flat base and others that are smaller.

2.4 Aff (Acromyrmex (Moellerius) striatus) ichnogenus. The broad distribution of this Attini encompasses Southern Brazil, Uruguay, Bolivia and Argentina as far as northern Patagonia. It is a species of field restricted to grassland, consuming Gramineae and dicotyledons. The population activity keeps free the surrounding of the nest entrance, lacking tumulus.

2.5 Aff (Pogonomyrmex bruchi) ichnogenus. This is an exclusively American genus, Pogonomyrmex is substantially xerophylus, and the greatest species differentiation is in arid environments, where they are all terrestrial. The genus is characterized by the little density of populations and colonies separated by vast spaces. They feed on plant seeds, to which they remove the glumellae, collecting them around nests and storing them in underground chambers. Some species are at the same time granivorous and hunters. A P. bruchi nest, made in solid or hard ground is superficial, has not a crater, the entrance is narrow, and in less compact soils, it has widened and circular longer passages. It differs from others for its more agglomerated chambers, distributed in levels for the breeds and seed accumulation, communicated by short horizontal passages. The nests host colonies up to 300 workers. Gatherer ants that include seeds in their diet have a significant impact on vegetation and their influence is not negative. The seed preference by the action of granivorous ants regulates the vegetable populations and their dispersion by myrmecocoria, a phenomenon that impulses the coevolution of ants and plants. This attitude allows recomposing and balancing the plant species distribution in some desert zones. In South America there is a marked absence of granivorous rodents, without competition with small ant colonies (Wilson and Holldobler 1990, Mares and Rosenzweig 1978).

2.6   The ichnofossils attributed to mound-building termites are relatively frequent in different levels of paleosols of Pampasia. They were mentioned for the first time by Laza (1995), who related them with the extant genera. At the upper Chapadmalal stage levels, several ichnofossils of ichnogenus aff (Procornitermes) were found. Genus formed by six species inhabiting the southern zone of the Amazon river basin, from which P. striatus occupies the most northern sector of the area, encompassing south of the biogeographic Chaqueña Province, north of La Pampa Province and the east edge of the Paranaense Province. Typifying ecologically such genus, Emerson (1952, p. 487) stated that "Procornitermes indicates by its distribution that it is a genus absent from the rainforest and typical of periodically drier savanna region."

3.5   Marplatan

3.5.1   Barrancalobian Sub-stage (Fig. 3.5)

This sub-stage coincides spatially with the Barranca de los Lobos "Formation" (Kraglievich 1952). By that time, marsupials record taxonomic diversity reduction, maybe due to the arrival of carnivorous forms from the Holarctic region such as mustelids, felids and canids (Pascual and Bond 1986). The presence of tayasuids and camelids led Menegaz and Ortiz Jaureguizar (1995) to relate them with open environments. Among Notoungulates, the genus Toxodon should be linked to water bodies (Bond 1999). The Xenarthra representatives record genera that persist until today, with taxa that suggest subtropical habitats, while the non-arboreal "Pilosa" suggests semi-arid grasslands (Carlini and Scillato-Yané 1999).

Oliveira (1999) mentioned the Marco Portugués site for the southern region of Brazil with the presence of hairy Megalonychops (present in the "Mesopotamiense" and Pliocene–Pleistocene of Argentina and Miocene–Pliocene of Uruguay) associated to large turtles, suggesting the prevalence of high temperatures.

Fig. 3.5 Marplatan—Barrancalobian sub-stage—Barranca de Los Lobos marine cliffs (photo by EP Tonni)

The ichnological record includes taxa corresponding to Coprinisphaera ichnofacies.

1. Numerous crotovines, some of large size, attributed to edentates, were measured by Zárate et al. (1998), counting up to 42 specimens
   The Krausichnidae ichnofamily is represented by termite nests:

   a. Tacuruichnus farinai (Cornitermes cumulans). This is a genus that inhabits tropical forests and savannas (Emerson 1952). C. cumulans lives to the south of the Amazon region, in the basins of the Paraguay and Paraná rivers, from 15° up to 30° in southern Brazil and center-north of the Chaco Province in Argentina, a region with savanna vegetation and tree patches. The ecological typification of this termite genus points out that the whole 10 species inhabit territories with rain regime higher than 1500 mm yearly (Constantino 1999) and average temperatures higher than 21 °C. Silvestri (1903), when describing the habitat of such species, stated: "The habitat of C. cumulans is dry with little or low arboreal vegetation." C. cumulans develops nests from a first hypogeum stage for then gaining land surface. The mounds, conical and irregular, reach 1.0 m high and the same size of circumference. The surface, mamellated, has rounded openings. The inner part of the mound is formed by an outer layer and nucleus, of different textures; the cells appear polished and covered with dark impregnations. Cornitermes cumulans present the most common nest in pasture and savannas of central and southern Brazil, as well

as in Paraguay and Argentina (Araujo 1970). Redford (1984) counted populations of 55 nests per hectare in the Brazilian region of Cerrado, where
savanna fires do not mainly affect its populations. In open grassland zones,
C. cumulans nests serve as agglutination of several organisms, becoming
a key species of the biota of those zones, such as ants and other termites.
Mound breakage and cracks host several uninvited guests: opilions, scor-
pions, spiders, myriapods, bees, wasps and, among vertebrates, mice, lizards,
snakes and several birds that became establish in the holes left by armadillos
and anteaters attacks. The tubular constructions express, from the environmental point of view, that they correspond to housing structures that during
part of their existence have to face flooding phenomena, as well as periodical
fires. Such structures, generally called "tacurúes" are made by Camponotus
ants and Cornitermes termites and even when they show outer similarity, the
inner organization is very different.

b. The Coprinisphaeridae ichnofamily is represented by C. murguiai specimens.

3.6 Marplatan

3.6.1 Vorohuan Sub-stage (Fig. 3.6)

In sediments of this sub-stage, there are three new taxa coming from North America:
Canidae, Equidae and Mustelidae. This sub-stage coincides spatially with the so-
called Vorohue "Formation" (Kraglievich 1952). Gasparini et al. (1986) pointed out
the presence of Tupinambis teguixin and worm lizard fossils to the east of Punta
Negra (Necochea County), while Tonni and Noriega (1996) mentioned the psittacid
Nandayus, a genus currently nesting in xerophytic forests of non-floodable areas of El
Pantanal in the high basin of the Paraguay River. Extant cingulate Xenarthra appear
related to subtropical conditions, maybe hotter than before (Carlini and Scillato-Yané
1999).

In addition, several Carnivora taxa were found: Stipanicia and Galictis ferrets;
Canidae represented by Canis, Ducicyon and Protocyon and Felidae by Smylodon,
the Pampas cat Felis Colocolo and a larger form related to Cougar.

The ichnolonogical record mentions several examples, all corresponding to the
Coprinisphaera Ichnofacies: (a) smooth tubes and meniscal holes denoting the
activity of oligochaete worms at paleosoil levels; (b) crotovines up to 1 m in diameter, as a result of the activity of rodents and large eutatines. The Krausichnidae
ichnofamily is also present through Nasutitermitinae or Termitinae (Barberichnus
bonaerensis) mound-building termites.

Fig. 3.6 Marplatan—Vorohuan sub-stage—Santa Isabel marine cliffs (photo by EP Tonni)

3.7 Marplatan

3.7.1 Sanandresian Sub-stage

This sub-stage coincides spatially with the San Andrés "Formation" by (Kraglievich 1952).

The correlation of several stratigraphic levels of Necochea and General Alvarado counties led Tonni et al. (1995) to conclude that sectors A and B of Las Grutas-Punta Negra and the sector B of Punta Hermengo correspond to this sub-stage. At its upper levels, geomorphological changes are noticeable related to the drop of sea level as a consequence of the glaciations under process; channeled sediments and the presence of strong diamictitic sediment banks are frequent. This climatic event seems to coincide with the first important cooling of the Cenozoic. Thus, the sub-stage may represent the moment of global climatic deterioration taking place around 2.5 Ma (Cione and Tonni 1995b). The fossil record points out the beginning of an important change in mammal associations for the upper part of this sub-stage, where markers of colder and more arid conditions than before appear Lestodelphys and the first large Tardigrade, as well as the presence of Gomphotheridae (a taxon from North America). At the same time, the last inhabitants of the hotter and wetter areas are recorded (Echimyidae and Tayassuidae). Carlini and Scillato-Yané (1999) coincided with this statement, affirming that Xenarthra suggest environmental conditions similar to the previous ones, maybe somewhat colder, while Pardiñas (1999), when studying sygmodontins, underlines that the high micro-herviborous diversity maybe related to colder and drier climatic conditions than nowadays. Verzi (1998) and Verzi and Quintana (2005), in turn, indicated that the fauna of fossil caviomorphs is a marker of aridity conditions.

The ichnological findings, corresponding to Coprinisphaera ichnofacies are:

1. Crotovines, some of large size, which Frenguelli (1921–1928) indicated as the largest of all outcrops of the coast of the Buenos Aire Province, stating that they corresponded to levels that at present are recognized as belonging to the Marplatan and Ensenadan Stages. The insect activity in paleosols is represented by two ichnofamilies:

(a) Krausichnidae, represented by mound-building termites. (Barberichnus) (Termitinae–Amitermes or Termes) at "Level A" of Punta Negra, Necochea County (Tonni et al. 1995). (b) Coprinisphaeridae, evidenced by numerous beetle balls (C. murguiaia and C. akatanka) (Cantil et al. 2013) at "Level A" of Punta Negra, Necochea County (Tonni et al. 1995).

In fact, this prolonged evolution moment of the biota in Pampasia, through numerous mentioned examples, verified the statement that during the Pan-Araucarian cycle several environments of the Chaqueño type, with some fluctuations, followed. Maybe one of the significant elements in all stages is the one involving all the elements: soils, vegetation, fauna and climate. It is "slag" and "terra cotta," which appear with certain profusion in diverse paleosols of all floors that form the cycle, as clear evidence of the development of savanna environments.

3.8   Faunistic Associations and Climatic Events During the Pan-Pampean Cycle

Even though the temporal duration of this cycle is significantly shorter than the previous one, the climatic events and its biotic correlatives were dramatic. After the first glaciation episodes detected in the upper levels of the Sanandresense substage, sediments corresponding to the Ensenadan stage began to accumulate. The ecosystem installation of Pampasia is recent, as the area was the development field of others that advanced on the region, with forms similar those surrounding it in the south, west and north. Process developed several times with different intensity. Since the middle Pleistocene (the Bonaerian stage) the Chaqueña influence weakened.

3.8.1   Ensenadan (Fig. 3.7a, b)

This stage coincides spatially with the Ensenada Formation in the northeastern Pampean region (Tonni et al. 1999a, b) and with the Miramar "Formation"

(Kraglievich 1952) in the Buenos Aires Province Atlantic Ocean coast. The correlation of different stratigraphic units in the Atlantic Ocean coastal outcrops concludes that the sector "C" of Las Grutas, Necochea County, as well as the middle areas of Punta Hermengo and basal areas of Costa Bonita in General Alvarado County belong to this stage (Tonni et al. 1995). Despite the biota degradation and impoverishment in the Pampean scenario, the Brazilian fauna extended its influence as far as Sierra de la Ventana and the city of Bahía Blanca. This fauna populated the region as far as its ecotonal borders to the west, in the garland of Sub-Andean and peri-Pampean hills bordering such plain like an open arc to the east. Ringuelet (1978) called that region the "Pampásico Domain," stating that it is a giant transition zone, a huge ecotone, whose historical analysis allows listing: (1) earlier extension of Brazilian fauna; (2) secular retraction of ecological causality; (3) advances and retreats of meso-Andean and southern faunas; (4) unstable balances and changeable ecotones; (5) bradytelic changes (slow) and tachytelic (fast) changes. Morrone (2000–2001) characterized the Pampean Province as a savanna area—sometimes floodable—with gramineae that can reach one meter high, herbs and bushes; xeric forests, similar to those in the Chaco Province, but impoverished, with forests in gallery along rivers. The long hydrographic basin from the Brazilian mountain range and represented at present by the Parano–Platense system developed progressively from the Miocene Sea retreat. It is known that the origin of the La Plata River began in Quaternary times, varying its width and depth according to the climatic cycles of that period (González Bonorino 1965, Parker et al. 1994, Cavallotto 1995). In the very long Ensenadan period, important faunal changes are verified, driven by the glacial advances and retreats, as well as the massive immigration of Holarctic mammals with a significant percentage of first records of genera and species (Tapiridae, Ursidae, Felidae, Cervidae) and among those autochthonous, a great diversification of xenarthra and megamammals.

The Ensenadan stage begins with the deposit of sediments forming the Mesotherium cristatum biozone. In the previous Jaramillo event, where hotter and probably wetter conditions than the present ones developed, the vertebrate fauna recorded in several sites of Pampasia could also be related to one previous episode to the Pleistocene glaciations. Among the varied available records, it should be mentioned that two magnetostratigraphic profiles, one in the outskirts of the city of La Plata and another in Buenos Aires City, are both carriers of Mesotherium cristatum. That of the city of Buenos Aires is associated to Tapirus and procyonidae Brachynasua, whereas remains of giant Geochelone and worm lizards were found in the Olivos locality (Gasparini et al. 1986). Besides, large forms such as Macraucheniopsis ensenadensis—recorded only in Buenos Aires City—and the tayassuidae Catagonus confirm the advanced mesic conditions (Bond 1999), a phenomenon repeated in Mar del Plata County with the finding of Mesotherium, Tapirus, Akodon cursor (?) and Nectomys squamipenis, which could indicate the broadest northern expansion of the Brazilian fauna at that time (Tonni and Cione 1995).

Sigmodontinae rodents verify the first record of extant species of the Pampean Region, whereas caviomorphs show extant genera already established. Vucetich et al. (1997) commented the finding of remains, similar equimid Clyomys (the most fossorial genus of the family) in Las Grutas-Punta Negra of Necochea County and

(a)

(b)

Fig. 3.7 Ensenadan and Bonarian—**a** northern of Mar del Plata marine cliffs; **b** "toscas del río de La Plata," at Anchorena station (photo by EP Tonni)

Mar Chiquita sites. They live in savannas in central-eastern Brazil and Paraguay, together with Plesiaguti totoi, Dasyproctidae whose two genera inhabit forests of Chaco and Cerrado (Vucetich and Verzi 2002). Another proof of fauna displacement is the finding of chelidae Hidromedusa near the city of Bahía Blanca, which moved its present habitat almost 500 km to the south (Deschamps 1995). Despite this, the prevailing conditions during the Ensenadan stage are extremely cold and arid. The mentioned avifauna for the Ensenadan stage indicates: Rheidae, Anatidae, Psittacidae, Oucudae, Phalacrocoracidae and passeriformes (Tonni 1980, Tambussi 1995). Among marsupials, the record is limited to scarce Didelphidae. Menegaz and Ortiz Jaureguizar (1995) mentioned the presence of tayassuidea, cervids (some of them extinct) and camelids. Nothrotheriums and megalonychids disappeared from the xenarthra Pampean fauna (only Megalonychops is recorded) and the presence of large armadillos and glyptodonts, as well as Tardigrade, a group that suggests steppes and open grasslands, maybe with tree-covered regions (Carlini and Scillato-Yané 1999). The Glyptodont Sclerocalyptus, with adaptations to arid and semi-arid environments as the development of front-nasal sinuses, a thermo-regulation mechanism that could have appeared in the Sanandresian-Ensenadan stages boundary (Zurita et al. 2005), is represented by four species. Both present genera of Pampatheriidae show adaptations to different environments: Holmesina, inhabitant of wetter zones and habitant of xeric environments (Scillato-Yané et al. 2005). A large presence of carnivores is recorded: skunks (Conepatus), ferrets (Galictis, Lyncodon) and otters (Lutra), as well as bears (Arctodus). Canids appear represented by robust forms (the forest species Cerdocyon) and small forms related with Pampean foxes (Ducisyon), whereas Felidae were represented by Smilodon and forms related to Puma concolor (Bond 1986)

In the Late Ensenadan stage, mammals indicating more arid and colder conditions are recorded, a moment that can be correlated to the beginning of the European glacial Pleistocene (0.80–0.50 Ma; Tonni and Cione 1995). Two findings are example of that moment and correspond to owl regurgitations, one in a crotovine to the southwest of Punta Hermengo, General Alvarado County, and the other in the sea gullies to the north of Mar del Plata City, General Pueyrredón County. The taxa that integrate both findings have taxonomic similarities with micromammals and birds of Mendoza forest and denote more arid and colder conditions than the present ones in the region (Tonni et al. 1993–1998). It is common to both findings the rodent Tympanoctomys, genus adapted to extreme xeric conditions, that develops isolated populations and semi-underground habits. Its migration is related with a glacial phase of the Pichileufú Drift, the sediments corresponding to the Great Patagonian Glaciation (the GPG; ca. 1.0 Ma, Early Pleistocene) in the Cordilleran zone (Verzi et al. 2002).

Ubilla and Perea (1999), when considering the Quaternary formations of Uruguay, assigned Ensenadan age to the Libertad or Raigón Formation and Preciozzi et al. (1985) assigned it to semi-arid climate conditions, according to sedimentological studies.

Oliveira (1999) indicated the presence of fauna that suggested conditions of mildly cold climate for some sites in southern Brazil, such as Paso do Megatério. The ichnites found in the Ensenadan stage correspond to Coprinisphaera ichnofacies.

1. Crotovines. In some outcrops on the Paraná River and in the town of Miramar, the latter built by rodents (Tympanoctomys?). To the north of Mar del Plata, in the locality of Vivoratá, remains were found in a cave of the bear Arctotherium angustidens with two cubs (Soibelzon 2009).
2. Regurgitations. To the south of Miramar and north of Mar del Plata city. The insect activity in paleosols is recorded by two families:

a. Krausichnidae, represented by two types of anthills:

(1) Ichnogenus aff (Pheidole spininodis). The Neotropical region is the region of largest differentiation of the genus Pheidole, reaching as far as 45° S and the Cordilleran forests in the parallels 38°–39° S. In Buenos Aires Province, it is the prevailing genus together with Acromyrmex, Solenopsis and Camponotus.

Pheidole spininodis, of great adaptive plasticity, is represented in arid and wet regions covering a large part of Argentina, Paraguay, Uruguay, southern Brazil and parts of Bolivia. It evolved from wetter environments to drier ones with the formation of a granivorous type (Kusnezov 1951). Anthill density and size are increased west of the Chaco, as a response to an increase of seed production caused mostly by annual species (in turn, resulting from the increasing aridity). Ants are the main seed predators in Chaco. It is noteworthy the low diversity of Chaqueña granivorous ant fauna, as it occurs in the forest.

It nests in solid lands, building a low and extended crater with wide circular entrance, making it easier the plant material transport to workers. The descending entrance channel, with endless twists, is a design that allows temperature and humidity regulation in the chambers; these—ranging from 5 to 10—, are small, low and vaulted, built overlapped around a vertical axis and connected by canaliculus. The nest depth exceeds a meter and a half.

(1) Ichnogenus aff (Pogonomyrmex bruchi) granivorous ant, inhabitant of semi-desert zones in western Argentina, with underground nests developed in solid soils. The association of these elements is interpreted as the establishment of colder and more xeric conditions than the present ones.

(b) Coprinisphaeridae, present through genus Quirogaichnus, found under a glyptodont shell in Entre Ríos Province (Frenguelli 1938a).

3.8.2   Marine Transgressions in the Pleistocene

The fluctuating line that relates the marine environment and its influence in the continent causes a populated area of continental organisms, permanently or temporarily, whose tracks are recorded in Scoyenia ichnofacies. Its domain and characteristic environments correspond to flooding plains, ponds, lake shores and ephemeral water bodies, and it shows a varied set of invertebrate traces and plants. The marine transgression of 6.5 million years ago, next to the Matuyama/Gauss border, was suggested as a boundary to define the beginning of the Quaternary chronostratigraphic subsystem (Pillans and Naish 2004). After 2.6 million years, the eustatic oscillations became strongly influenced by the glacial–interglacial cycles that characterized the Pleistocene (Buchman et al. 2009).

Along the Atlantic Ocean coast of this region, at the mouth of some rivers and streams, deposits of marine origin are sparingly recorded. This evidence corresponds to diverse marine transgressions of the Quaternary (Schnack et al. 2005). Between 123,000 and 18,000 years ago—the last full glacial period—, sea-level fluctuated, descending up to 100–120 m below present sea level, or invading the continent at more restricted levels in all cases. The last transgression began 7000 years B.P. and between 5000 and 6000 years B.P., it reached its maximum level followed by a regression up to the present level. As result of these phenomena, different sedimentary deposits with own features and fossil evidence—mostly the malacofauna–, allowed radiocarbon dating and paleoenvironmental studies.

The marine transgressions recorded in the Pampasia region are:

1. "Inter-ensenadense." Dated at 2.4 Ma, it was much older than the Last Interglacial (Schnack et al. 2005).
2. "Belgranense": this transgression was considered as correlative with the Sangamon epoch or the Illinois–Wisconsin interglacial period by Pardiñas et al. (1996). Chronologically situated in the sub-stage 5e (ca. 120,000 years ago), its deposits appear restricted and discontinuous along the entire littoral region, represented by different facies, intercalating or covering Pampean sediments, with higher levels at 6–8 m above present sea level. Its stratotype is found in the Belgrano neighborhood, Buenos Aires city, and it is considered to be intercalated in the Ensenadan stage (Cione et al. 2002). Deposits appeared in Santa Clara del Mar, north of Mar del Plata City and in Pehuén-Có locality (lower section of the San José Sequence; Zavala and Quattrocchio 2001). Similar deposits are recorded in the Colonia and Rocha departments, Uruguay. However, the age of the "Belgranense" units is not clear, and if the accepted age is within the Ensenadan stage, it cannot be coeval with the Last Interglacial period.
3. The Pascua Formation. Recent mineralogical and geochronological studies placed stratigraphically such information intercalated between the Bonaerian and Ensenadan stages. The Pascua Formation was considered as "Belgranense" by its author (Francisco Fidalgo) and therefore assigned to the last interglacial (5e).

The system of littoral barriers developed in the Río Grande do Sul coast, Brazil, was developed in several stages, from which only four have been identified, assigned to interglacial stages 11, 9, 5 and 1, attributing them ages of 400,000, 325,000, 123,000 and 6000 years B.P., respectively. In Barrier I, mammal crotovines were found and in Barrier III, ichnofossils of digger crustaceans and mollusks were found in different localities at 7 m above present sea level.

3.8.3   Bonaerian (Fig. 3.8a, b and c)

This stage coincides with the Buenos Aires Formation in northeastern Buenos Aires Province (Tonni et al. 1999a, b; Nabel et al. 2000). The observations by Teruggi and Imbellone (1987) on the paleosols of that same age in the Gorina locality, at the outskirts of La Plata City, indicated that the hydromorphic features are more evident in lower levels, pointing out more arid conditions toward the present. The Bonaerian stage starts probably with a warmer event, pedogenesis prevalence and faunal record of the Chaqueño subtropical elements (Echimydae, Dasyproctidae, Noctilionidae) (Vucetich and Verzi 2002). Tentatively, the Bonaerian base is correlated with the Isotopic Stage (IS) 11 (ca. 0.40 Ma; Verzi et al. 2004), an event that defines a biozone characterized by the presence of the rodent Ctenomys kraglievichi. Its type area is composed of the outcrops at the Sauce Grande River, in Bajo San José, near Bahía Blanca County. Here, a vertebrate association was identified that includes fish of Brazilian and southern origin, some of which do not inhabit the zone at present, as well as turtles, marsupials and rodents inhabiting the place during a quite marked warmer pulse (Cione and Báez 2007). Such fauna contrasts suggest the existence of fast and great environmental changes that reflect individualistic responses of the involved species. The biozone is recognized also in Necochea County (Costa Bonita, Las Grutas-Punta Negra) and to the northeast of Mar del Plata City (Camet and Constitución neighborhoods). Ctenomys kraglievichi appears related with the living Clyomys, of close similarities with the Ctenomys from Bolivia and Paraguay, which inhabit savannas and forests of Cerrado in Brazil and eastern Paraguay. The remains found in Santa Clara del Mar appear immediately on top of the Belgranense transgression, between the Ensenadan and Bonaerian stages, and it correspond to a possible interglacial period.

Alberdi and Prado (1995a, b) mentioned two different equid genera in Pampasia: Hippidion, inhabiting from the north as far as Buenos Aires Province on soft soils and warmer climate and Eqqus, inhabitant of colder areas as far as southern Patagonia. Meanwhile, Menegaz and Ortiz Jaureguizar (1995) pointed out the presence of camelids, cervids and tayassiuds as markers of open environments with forest patches. Oliveira (1999) mentioned that the fossil fauna found in Irai and Campo Seco localities in southern Brazil, indicates moderate climate conditions.

Then, the biozone of Megatherium americanum follows, a period during which xeric and colder conditions developed, increasing toward the end of this period.

In these levels, at the Buenos Aires Atlantic coast, crotovines of great size and some edaphic levels with meniscal tubes (Taenidium) are frequent. In the Centinela del Mar locality, in its upper levels, an anthill of Acromyrmex (Moellerius) striatus, representative of Attini, was found, corresponding to the group of conqueror ants of arid zones.

The presence of strong diamictitic levels in the area of Buenos Aires Atlantic coastal gullies indicates that during several moments the marine level underwent significant variations, some of them quite important. The loss of base level of the water courses descending from the Sierras of Buenos Aires Province, in their search for

Fig. 3.8   Bonarian—Tapalqué Creek (photo by EP Tonni)

balance, resulted in deep dug beds, then filled by important sediment accumulations, some in the shape of diamictons.

The ichnological findings in this stage correspond to Coprinisphaera ichnofacies:

1. Smooth tubes and meniscal holes, in outcrops on the Río Paraná gullies between Rosario and Campana localities (Voglino 1999).
2. Crotovines, some older ones, in the same sites as the previous one (Voglino 1999).

The insect activity in paleosols is represented by two ichnofamilies:

a. The Krausichnidae Ichnofamily, present through two types of termite mounds and an anthill.

   I.   Tacuruichnus (Cornitermes). Genus of savannas and tropical forests of South America (Emerson 1952). It was found in Centinela del Mar, General Pueyrredón county, at the base of the Bonaerian sediments.
   II. Barberichnus bonaerensis (Termitinae, Amitermes or Termes). Currently, the southern border of Termitinae, capable of building such niches, is 32° S latitude, a territory corresponding to the Chaqueña and Paranaense biogeographic provinces. The remains were found at the base of the Bonaerian beds in the city of La Plata (Fig. 3.9a, b)
   III. Ichnogenus aff (Acromyrmex (Moellerius) striatus). Found in Centinela del Mar, General Pueyrredón County, at the upper Bonaerian levels.

(a)

(b)

Fig. 3.9 Ensenadan and Bonarian—**a** and **b** excavations for the construction of the Teatro Argentino of La Plata (photos by EP Tonni)

b. Coprinisphaeridae Ichnofamily, represented by Scarabaeinae nest balls. It was found in General Pueyrredón county (Centinela del Mar, Santa Isabel Beach), associated to Lagostomus, Reithrodon, Dolichotis and Tayassu.

Verde (2000) pointed out the finding of Prosopis (carob tree) fossil wood with galleries attributable to cerambycid coleopteran in Uruguay, Río Negro Department, in the Dolores Formation.

3.8.4 Lujanian (Fig. 3.10)

Equus (Amerhippus neogaeus) biozone is the Lujanian biostratigraphic base, as it was defined by Cione and Tonni (1999–2001). Besides, the marine coastal levels of the Pascua Formation (Fidalgo et al. 1973a) correspond to this stage.

The Lujanian stage begins with interglacial conditions (Isotope Stage 5e, 130 ka B.P., the base of the Late Pleistocene). Its final part developed during the Last Maximum Glacial (18 ka B.P.) including the last glacial advance, equivalent either to the Younger Dryas period or the Antarctic Cold Reversal, or both (ca. 13–11 ka B.P.). During the Middle and Late Pleistocene, 20 couples of glacial/interglacial changes followed; these changes took place so fast that in some sites cool climate elements appear associated to others of warmer zones (non-analogous associations). The

Fig. 3.10 Lujanian and Platan—Quequén Salado River (photo by EP Tonni)

fauna adapted to arid and colder conditions prevailed in that period (Tonni et al. 1999a, b; 2003) and diverse camelids, among mammals, prevailed at moments of greater aridization. Such aridity conditions were detected in Uruguay (the Dolores Formation) and in Rio Grande do Sul, Brazil (the Santa Victoria Formation). This was the moment when the first Homo sapiens are recorded in South America as well as extinctions of endemic mammal families and Holartic taxa (Pampatheriidae, Glyptodontidae, Megatheriidae, Mylodontidae, Megalonychidae, Macraucheniidae, Toxodontidae, Gomphotheriidae, Equidae), whereas others restricted their living areas (Tapiridae, Tayassuidae, Ursidae). At the same time, mammals originated in the central and Patagonian areas were then detected in the Pampasia territory, together with plant elements from the Monte (Larrea) and Espinal (Prosopis) botanical provinces.

At the end of the Lujanian stage, two mammal associations were detected: one for La Chumbiada Member and the other for the Guerrero Member, whose faunal differences include the decrease in diversity and frequency, as well as the increase of mega-herbivore grazers, ecological succession interpreted in the frame of climatic deterioration. In the Guerrero Member, vertebrates indicating arid and colder conditions from Central and Patagonian areas were detected, such as the marsupial Lestodelphis and birds such as Pterocnemia (Tonni and Laza 1980a). The lower part of the La Postrera Formation, with a similar fauna to the Guerrero Member, includes the last mega-herbivores and the extant species of Central and Patagonian areas (Tonni and Cione 1995) in eolian sediments including forms of degraded sand hills. Both members were related with the Full Glacial Phase (Tonni and Fidalgo 1978); later, Tonni et al. (1999a, b) correlated the Guerrero member with the Last Glacial Maximum (LGM), which, according to several available dates, may have taken place between 21.0 and 10.0 ka B.P. Such period comprises 2/3 of Isotopic Stage 2 and beginnings of Stage 1, during which several climatic events occurred: around 20.0 ka B.P., the maximum glacial advance; from 16.0 ka to 14.0 ka, a fast increase of temperature (phasal) and around 11.0 ka B.P., new glacial advances took place (Antarctic Cold Reversal? Younger Dryas?), colder events which are not clearly identified yet in Pampasia. For additional information concerning the Pleistocene glaciations, see Rabassa (2008).

The record of an arid episode in 10.0 ka B.P. suggested warmer conditions (Pardiñas and Lezcano 1995). At the end of the Pleistocene (ca. 10.0 ka B.P.), warmer and wetter conditions generated a pedogenetic episode, represented by the Puesto Callejón Viejo "chernozoid" paleosol (Tonni et al. 2001), while the last extinction events took place 7000 years ago (Tonni et al. 2003).

Lujanian stage ichtiofauna was recorded in the Guerrero Member (Paso de Otero, Río Quequén Salado and Estancia La Moderna in the Azul County) (Cione and López Arbarello 1995).

When studying reptiles, Gasparini et al. (1986) pointed out the occurrence of Tupinambis teguixin near Buenos Aires Province, whereas Albino and Carlini (2007) mentioned the presence of Boa constrictor in sediments of the Toropi creek (Corrientes Province), a record that does not correspond to the pattern of present distribution. Sites situated in the central-northern Santa Fe, Corrientes, and Entre Ríos provinces, northern Uruguay and southern Brazil, appear as favored by more benign environmental conditions, with faunal elements common to many of them, such as browser notroterine Neolicaphrium and giant turtles. Apparently, in those areas the warmer conditions lasted longer, with prevalence of Brazilian flora and fauna. Noriega et al. (2004) mentioned a singular group of vertebrates for the Ensenada creek in Entre Ríos province. Glyptodon perforatus, giant turtles Emydidae and Testudinidae, pampaterium Holmesina paulacoutoi, Tapirus and the giant otter Pteronura, are Brazilian elements that show the development of a biogeographic area independent of that Central-Pampean area, at a moment of interglacial conditions that may correspond to the Isotopic Stage 5e (ca. 13.0 ka B.P.) (see also Ferrero 2007). Similar position is stated by De La Fuente (1999) when mentioning the large turtles Chelonoidis in the Carcarañá and Salado rivers in Santa Fe Province and by Tonni (1992a, b) when pointing out the presence of Tapirus terrestris, giant turtles and Hidrochoerus in the Perucho Berna creek, in Entre Ríos province. When referring to ornitofauna, Tonni (1980) and Tambussi (1995) cited Tinamidae, Rheidae, Rallidae, Strigidae, Phalacrocoracidae, Ciconiidae, Charadriidae and Cathartidae, all belonging to extant families and genera, except for the Tinamidae.

Caviomorph rodents underwent slight changes in the distribution in the last 10.0 ka B.P., while Sigmodontins decreased in diversity, showing corological changes of subtropical and Patagonian-central taxa (Pardiñas et al. 1996, Vucetich and Verzi 1999).

Menegaz and Ortiz Jaureguizar (1995) mentioned several camelid genera, while giant Xenarthra showed more noticeable xeric conditions.

Carnivores appeared quite diversified: skunks, mustelids, otters, bears of Arctodus genus, diverse foxes and large canids (most of them became extinct during this time), whereas Felidae appeared through extinct and extant forms such as the jaguars.

Many deposits of Late Pleistocene age were recorded: the Santa Clara Formation in Camet Norte, north of Mar del Plata City, Mar Chiquita County (Fasano et al. 1984), and others in the area of Punta Hermengo (Tonni and Fidalgo 1982), whose fossils form "non-analogous associations." In Punta Hermengo, continental fish, amphibians, reptiles and mammals of Brazilian lineage were found, together with Cavia of Central origin, showing warmer and wetter episodes (Pardiñas et al. 2004). Warmer pulsations were recorded in Punta Hermengo and Camet Norte from where Noriega and Areta (2005) described the first Sarcoramphus papa record, whose present distribution border places it 700 km to the north, associated to Holochilus brasiliensis and Myocastor sp (representatives of the Subtropical Domain sensu Ringuelet (1961). Such findings suggested the existence of a plant community able to support such taxa. In Brazil, the Touro Paso Formation is carrier of similar elements. Such evidence is opposed to ideas expressed by Pardiñas et al. (1998), who claimed climatic homogeneity dominated by aridity.

The most modern units of the Pleistocene appearing in Mesopotamia correspond to the Lujanian stage, explaining discrepancies regarding the Pampean biostratigraphic scheme as the result of differential biochrones of taxa in different areas (Tonni 2009). The new systematic hypotheses show that the northern sector of Mesopotamia had greater fauna links with eastern Brazil than with the Pampean region (Carlini et al. 2004).

Ubilla and Perea (1999), when studying the Sopas Formation in Uruguay, informed that this is carrier of subtropical to tropical elements (Tapirus, Catagonus, Hydrochoerus, Erethizonthidae), including mollusks and paleosols with trunks in life position and ichnofossils (Verde et al. 2006), suggesting a non-classical biogeographic pattern. Datings assign it an older age than the Guerrero Member of the Luján Formation. The Sopas Formation was correlated with the Feliciano Creek Formation in Entre Ríos Province with the Formation Touro Passo in Brazil, by Anton (1975) and Bombín (1976), whereas Oliveira (1999) added that this involves fauna of wetter environments and weaker winters, correlating it with the Barranca Grande Formation. The record of proteroterid Neolicaphrium, marginal element of the Pampa region during that time, appeared confined to warmer forested zones (Bond et al. 2001). Also, the Toropí Formation, analyzed by OSL techniques, yielded ages between 50 and 35 ka ago. The biochronological analysis of these units showed greater links with eastern Brazil (k et al. 2005), whereas fossils of the Dolores Formation in Uruguay indicated Pampean-Brazilian affinities in arid environmental conditions, able to be correlated with the Guerrero Member of the Luján Formation. Oliveira (1992) mentioned a similar fauna for the Santa Victoria Formation in Río Grande County.

At Osorio, in the coastal area of Brazil, there is the Chui Creek site composed of two sedimentary groups, the lowest one including Lujanian mammals in immature paleosols (eolianites barely fixed by vegetation). Insect nests were found, and two groups were distinguished:

(a) Termitichnus, composed of termite nests (Vondrichnus, Krausichnus and Termitichnus) and oligochaete holes (Taenidium barreti);
(b) Celliforma, with plenty of cells, Monesichnus and Palmiraichnus, cells of Uruguay, wasp cocoons and beetle nests (Coprinisphaera fontanai), Taenidium barreti and discret rhizoliths (Netto et al. 2007).

The higher levels give place to Skolitos ichnofacies, evidence of the sea advance and maybe it can be correlated to the last Pleistocene transgression (the "Querandinense" strata).

Paleophytology in the Pampasia area recognizes two large sedimentary environments:

(a) Sedimentation of different units that form the plain area, where the conditions have not allowed the preservation of vegetable remains, being recorded only in the Lujanian and Platan stages;
(b) The area of Mesopotamia, where the stratigraphic succession provided suitable conditions for the preservation of plant remains. Paleobotanic studies of the Pleistocene–Holocene units suggested that the vegetation changes were linked mainly to models of atmospheric circulation and to those of glaciations (Barreda et al. 2007).

Hydrophilic forests persist in Mesopotamia, recognizing two paleophytogeographic provinces for the region:

(a) Neotropical, with elements related to the present "Paranaense" Province and

(b) Yungas, with vegetation similar to the present Chaqueña Province, with xerophilus forests, palm groves, savannas and bushy steppes, halophytes and hydrophilic communities linked to water courses (Barreda et al. 2007). As an example, the paleocommunity that provided El Palmar Formation (Iriondo 1980) in the Colón Department, Entre Ríos Province, was cited. Dated at the interval Isotope Stage 5, the wettest and hottest of the Late Pleistocene (Iriondo and Krohlin 2001), this profile presented two levels, the upper one with Palmoxylon trunks in life position and phytoliths of panicoideae type, characterizing a savanna community, and the lower level with angiosperm remains and aquatic flora of fluvial origin (Zucol et al. 2005; Brea and Zucol 2007; Herbst et al. 2007). Tonni (1987) described Stegomastodon platensis remains from that formation, and when referring to Quaternary proboscidean, Alberdi and Prado (1995a, b) mentioned the presence of two genera of open environments: Cuvieronius of smaller size for the Early Pleistocene and Stegomastodon of larger height for the Middle and Late Pleistocene.

Paleobotanic studies in several zones of the Buenos Aires Province identified:

(a) Romero and Fernández (1981) described pollen associations in paleosols near Lobería City and core samples ok Lake Chascomús;

(b) Prieto (1996–2000), who carried out similar studies in several sites of Buenos Aires Province, such as Sauce Chico creek, Fortín Necochea, Empalme Querandíes, Cerro La China and Las Brusquitas creek. The sequence comprised Lujanian and Holocene beds and listed a succession of communities that showed significant environmental changes:

   (i)   Before 10,500 yr B.P., psammophytic vegetation similar to the current one with little rainfall prevailed;

   (ii) The coastal line, situated 150 km to the east, created greater continentalism and the hydrophytic communities changed to grasslands;

   (iii) During the period between 8000 and 6000 yr B.P., sea level rising imposed wetter climatic conditions, and the communities of psammophitic and holophytic vegetation returned. Quatrocchio et al. (1995) and Schabitz (2003) studied palynology in paleosols in the zone of the Sauce Grande River, the Napostá Grande Creek and southwestern Buenos Aires Province. According to these authors, the oldest sequence (Late Pleistocene) evolved from somewhat wetter eolian environments to arid climate and the plant communities corresponded to the invasion from the periphery of different communities such as the Monte Formation (with Larrea bushes) and the xerophytic forests of Espinal, with Prosopis trees and dense graminous coverage. The second association (beginning of the Holocene times) with wetter, milder climate to semi-arid conditions, and communities that changed from graminous steppes to herbaceous-psammophytic ones. Paleosols indicate the beginning of less cold, and wetter climatic phase for the Pleistocene–Holocene transition, and they become interfingered in the Napostá Grande Creek mouth with marine facies correlated with the Hypsithermal times (Middle Holocene). The current grasslands are divided in "Pampean Grasslands" of Argentina and "Campos" of Uruguay and southern Brazil. The "Pampean Grasslands" are surrounded by a xerophytic forest that extends in the shape of an arch in the western, northern and northeastern sectors.

The ichnological findings in this stage correspond to two ichnofacies:

(a) Scoyenia ichnofacies with the following elements:

   1. Taenidium meniscal holes in Pehuén-Có
   2. Vertebrate footprints in Pehuén-Có, Monte Hermoso, Sierras de Tandil coves and Santa Clara del Mar, Buenos Aires Province.

Coprinisphaera ichnofacies appears with the following elements:

(i)   Coprolites in northern Mar del Plata County.
(ii)   Gastrolites in the Luján River Basin.
(iii) Owl regurgitations in the lower levels of Cueva Tixi (Sierras de Tandil).
(iv)   Crotovines along the Atlantic Ocean coast of Buenos Aires and Uruguay.
(v)   Taenidium serpentinum meniscal holes and Castrichnus incolumis estivation chambers in Uruguay, and Edaphichnium tubes in La Plata County.
(vi)   Insect marks on bones in deposits of the Río Quequén Grande (Pomi and Tonni (2010).

The insect activity in paleosols is represented by two families:

(a) The Krausichnidae family, represented by three types of anthills and one termite mound:

   1. Aff Acromyrmex (Acromyrmex) lundi ichnogenus: its distribution covers the Buenos Aires Province surroundings as far as the Río Negro Province, being mentioned in Uruguay, Paraguay and Bolivia. Field ant species, dicotyledon eater, whose nests often develop at the foot of trees. The anthill has a low tumulus (40–50 cm) and 2 m in diameter at the base; the underground section, with several passages leading to a unique large fungus accumulation (locally known as "hongueras") (more than 0.50 cm in diameter) at 1–2 m deep. The specimen was found in cuts of tributary gullies of the Río Quequén Grande, in Necochea County, Buenos Aires Province.
   2. Aff Forelius chalybaeus ichnogenus. They form small ant colonies, terrestrial, which inhabit arid, semi-arid zones and xerophytic forests. They form part of the core of arid zone fauna whose species have evolved as autochtonous elements of the zone (Kusnezov 1963).
       F. chalybaeus inhabits the Pre-Cordillera arid strip and northern Patagonia. It forages cactaceous fruits and hunts insects. It builds nests in places free of vegetation, in sandy and solid soils, and when loose, the chamber walls

are protected by a resistant crust. It builds low craters, of regular shape (10–20 cm in diameter) with a small entrance hole; a short vertical passage leads to irregular chambers, much longer than they are higher, communicated by extremely narrow passages. The total nest depth is around 20 cm.

Several nests were found in coastal gullies to the north and south of Mar del Plata County, Buenos Aires Province

3. Aff. Trachymyrmex ichnogenus, member of the Attini tribe founded by Frenguelli on the banks of the Río Salado Norte, in Santa Fe Province, from which molds of two fungus accumulations ("hongueras") were recovered, and they are kept in the collections of the La Plata Museum.
4. Tacuruichnus sp. Termite mounds found in the zone of the Toropi Creek, Corrientes Province (Erra et al. 2016).

(b) Coprinisphaeridae family, represented by two Coprinisphaera species:

1. C. murguiai, a nest associated to Stegomastodon remains found in the Río Salado, Las Flores County, Buenos Aires Province.
2. isp. A (sensu Sánchez 2009), several specimens from the Venado creek to the west of Buenos Aires Province.

Other possible Coprinisphaera remains were mentioned for southern Santa Fe Province, on Paraná gullies between the cities of Rosario and Buenos Aires and the Corrientes Province. Finally put an end to them, when penetrating these territories around 13,000 years B.P.

Megamammals (36 species of 18 genera) and large mammals (46 species of 30 genera), which were present in the Lujanian stage, became extinct (Cione et al. 2007).

The disappearance of mega-herbivores could result from the plant impoverishment in many areas, an event that caused the isolation of several populations of this fauna.

3.8.5 Pleistocene Extinctions and the Arrival of Man

The last and most important fauna changes in the region took place during the Late Pleistocene–Holocene, when about 80% of the large mammal species, herbivores and carnivores became extinct in coincidence with the arrival of humans to the region.

Cione et al. (2003, 2007) claimed that, during the Late Pleistocene, the successive and severe climate changes caused dramatic modifications in the biomass of South America and other continents, and the phenomenon called "Zig-Zag Roto" took place. During this episode, the mega-herviborous fauna could not adapt to the frequent climatic changes and consequent plant changes, which produced the impoverishment of populations. Humans had difficulties in feeding and procreation. Darwin (1844) had observed and written in his essay that … "animals, apparently, become sterile when being taken out from their natural condition more often than plants"… It was during this moment when the populations were more vulnerable, that human hunters could have led most part of those populations to extinction.

Archeological sites are recorded in Pampasia at the end of the Pleistocene-Holocene transition, but ichnite carriers are scarce:

1. Late Pleistocene–Early Holocene (12,000–6500 years B.P.), represented in the Pampean plain by the Site 2 of Arroyo Seco, Tres Arroyos County, in the Buenos Aires Province central area, between the two hilly ranges of Tandilia and Ventania. The extinct fauna elements found, characteristic of the "Lujanense," include Megatherium, Mylodon, Glossotherium, Equus, Hippidion, Toxodon, Macrauchenia and Hemiauchenia (Fidalgo et al. 1986). Some of their remains present signs of picking and defleshing. One of the several human burials of the site included as part of the funerary garments a Glyptodon plate.

The lower stratigraphic level of the site was correlated with the arid phase when La Postrera Formation was deposited (Fidalgo and Tonni 1982). Its "S" layer corresponds to an edaphic horizon, from which several Coprinisphaera murguiai specimens were extracted. The nests are similar to those of canthonino Megathopa villosa, Scarabaeinae genus currently represented in the Patagonian region and Chile, elements that indicate an advance of Patagonian and "Monte" faunas over the Pampean zone.

Then the extended pedogenetic episode known as the "Suelo Puesto Viejo" took place in Pampasia, followed by a long arid phase until the Late Holocene and represented by Central and Patagonian mammal species. An example of this transition of the Pleistocene–Holocene times and of the fauna displacements, are the deposits of the lower level of Cueva Taxi—situated in the Sierras de Tandil area—represented by mammals that included extinct and Patagonian elements. Deposits that conclude with pellet layers, carrier of Scarabaeinae nests Onthophagus hírculus, which occupied the interior of the cave, compelled by the external climatic conditions and the food supply inside it (Laza 2001). Such pellet layer, together with the bat excrement deposits was populated as "soil" by the mentioned Scarabaeinae. The fauna group of the upper level corresponds to transitional conditions from arid to mild-wetter, has fauna with Brazilian influence and was dated in 6500 years B.P.; it was followed by a new glacial expansion that left as evidence the existence of a second pellet layer with nests of the mentioned beetles.

On the Atlantic Ocean coast of the inter-hill area of Buenos Aires Province, there is the ichnological site Monte Hermoso I, 6 km to the west of Monte Hermoso locality. This site is carrier of numerous human footprints, birds and an artiodactyl. These are printed on sediments of the Early Holocene, appear discontinuously distributed along 800 m on the present coastline (Politis 1984, 1993). The carrier layer, composed by alternating clays and sand, corresponded to an interdune, shallow lake body. The deposit provided Ruppia seeds and bony remains of the sea wolf Arctocephalus, with ages of 7125 ± 75 and 7030 ± 100 years B.P. A large part of the footprints correspond to children and younger adults, and they may be linked with the archeological site "La Olla," nearby, of sea wolf processing and consumption, dated in 6640 ± 90 and 7315 ± 90 yr B.P. and 7315 ± 75 yr BP (Politis 1993).

3.8.6 Holocene Marine Ingressions (Fig. 3.11)

The marine ingressions of the Holocene left deposits with levels up to 5 m above present sea level or lower, when penetrating in the Río de la Plata system as far as southern Entre Ríos, covering low coastal zones in Pampasia area. Its study allowed relating sedimentary environments with postglacial changes of sea level (Violante and Parker 2004).

Registered ingressions are represented by:

1. Destacamento Río Salado Formation ("Querandinense")
   The maximum elevation of this transgression was around 6000 years B.P, and the dropping of sea level has been evident from 3000 years B.P. to present times. Deposits on the Atlantic Ocean coast showed good expositions formed by coastal lagoon, sandbars and beaches. It affected coasts of Buenos Aires, southern Entre Ríos and Santa Fe Provinces, Uruguay and southern Brazil; crab habitats developed in some of these environments.
   In Arroyo Chui site, the Osorio locality in the southern littoral coast of Brazil, Netto et al. (2007) distinguished sediments carriers of ichnofossils corresponding to Skolitos (sic) ichnofacies that can be attributed to the "Querandinense" sea.

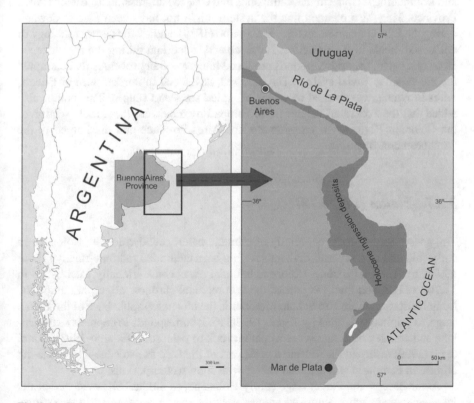

Fig. 3.11 Holocene marine ingression. Modified from Tonni 2017; drawing by Marcela Tomeo

2. Las Escobas Formation, formed by two members: Canal 18 and Cerro de La Gloria members, also known as the "Platense" (Fidalgo et al. 1973a). It formed two vast coastal barriers, the first one from the Río Salado to the Mar Chiquita lagoon, the second from Punta Hermengo to Pehuén-Có, with a maximum width of 3.5 km (Isla et al. 1996).

Similar marine fluctuations were detected at the coast of Rio Grande do Sul State, Brazil, in the shape of alluvial fan and barrier lagoon deposits (Villwock and Tomazelli 1996). The study of the landscape evolution in plains is only allowed by the interpretation of surface geoforms, just like the study of the hydrological system evolution in Pampasia during the Late Pleistocene–Holocene. Many researchers coincide on the fact that the conditions of water streams varied or underwent alterations during that period with respect to their present development. Iriondo and Krohling (1996) pointed out that during that period, the hydrographic system Bermejo–Desaguadero–Salado rivers transported important flows during the wetter periods of the mentioned period, as well as the summer transport of their Cordilleran tributaries, with strong snow fields and glaciers in their headwaters. The main collector flowed from north to south over 1000 km and then to the east, excavating many valleys in the La Pampa Province (partly blocked by eolian sediments at present) and lengthening its trace in the Vallimanca and Chasicó depression, in Buenos Aires Province. They also claimed that the system might not have been able to dewater during the Last Glaciation. In turn, Malagnino (1989) studied the geomorphology of eolian sediments in Buenos Aires Province and stated that during the Pleistocene–Holocene cycle, the aridity periods produced intense eolian processes that disrupted the pre-existing fluvial system. The sand sea, developed in northeastern of Buenos Aires Province, was formed by dunes of varied sizes and shapes. This author also added that the excavated valleys in La Pampa Province were the middle section of the Colorado River when crossing Buenos Aires Province and discharged in the Samborombón Bay.

3.8.7 Platan (Fig. 3.12)

The generalized criteria, especially of climatic nature, subdivides the Holocene in early, mid and late, this sub-divisions having been calibrated radiometrically (Tonni 1992a, b). The Pleistocene–Holocene boundary was conventionally established in 11,700 yr B.P. and involves part of the two regional chronostratigraphic units: the Lujanian stage, whose Río Salado Member is the biostratigraphic base of the Platan stage, as originally defined by Tonni (1992a, b). It corresponds to deposits of fluvial-lake and eolian origin that serve as parent rock to paleosols (Fidalgo 1993). Tonni et al. (1999) affirmed that during a great part of the Late Pleistocene–Holocene the climate in Pampasia was arid and colder with short wetter periods.

At the base of the Platan stage (Early Holocene, ca 8.0 ka B.P.) evidence of the Patagonian and Central fauna was recorded in the region, verifying arid conditions,

Fig. 3.12 Platan—Miramar, Punta Hermengo (photo by EP Tonni)

whereas in ca. 7.0 ka B.P., a warmer and wetter event began, originating tropical fauna, pedogenesis and expansion. The climate continued fluctuating in the subsequent millennia forming the current Pampean ecosystem. Species of autochthonous fauna are exclusively recorded in this biozone (Lagostomus maximus), except for the canid Ducysion avus, which became extinct around 1600 years B.P. (Tonni and Politis 1982), and some species of the Pampean megamammals, whose more modern records are somewhat older than 7000 years B.P., such as Eutatus seguini, Doedicurus sp., Sclerocalyptus sp., Mylodon listai and Megatherium americanum.

Cione and López Arbarello (1995) mentioned that the ichtiofauna in the Río Luján and Ensenada localities was represented by Siluriformes and Characiformes, elements from the "Parano-Platense" biogeographical province.

Besides, Menegaz and Ortiz Jaureguizar (1995) pointed out the presence of Lama and small deers, whereas Pardiñas (1999) indicated that Sigmodontine rodents, with maximum specific diversity in "non-analogous" aggregates during the Late Holocene, allowed inferring greater warming in southwestern Buenos Aires Province.

During most of the Holocene, the prevailing elements corresponded to eremic conditions, that is, why the wetter episodes do not seem to have been intense or longer enough in order to impact the vertebrate biota. Finally, species that respond to wetter conditions (Vizcaíno and Bargo 1987) and arid fluctuations (Politis 1984) followed.

Some coastal sites represented accurately the environmental events of this period:

1. Tonni and Cione (1984) described a curious thanatocoenosis of the Las Escobas Formation in the boundary of the La Plata and Ensenada cities. Together with remains of Eubalaena australis, they found other remains of sharks, rays and teleosts and rodent and worm lizard waste. The mixture of different taxa was interpreted as a replacement episode of eremic fauna by a Brazilian one.

2. Rossi et al. (2001) described, in the Punta Hermengo profile:

   (a) Tubes with oxidation halo of 24 mm in diameter by 50–340 mm long, sharpened end and dark or meniscal massive filling with the concavity upwards, similar to ichnites with remains of Tagelus plebeius, of Holocene age, found in Mar Chiquita;

   (b) Horizontal tunnels, transgressive to estuary sediments, of 50–70 mm of section and up to 750 mm long with dark laminar filling converging in an ovoid chamber of 790 mm and were interpreted as vertebrate caves;

   (c) Tubular rootlets of 8–45 mm long and 10–50 mm in diameter almost along all the profile and at its base rhizoliths covered by iron oxides from 7 to 70 mm long and 0.5 to 10 mm in diameter.

3. Stutz et al. (1999) described a pollen profile in La Ballenera Creek, in sediments corresponding to the "Querandiense"-"Platense" beds that evidenced vegetation and environmental changes caused by marine oscillations: (a) 6790 yr B.P., of continental origin; (b) until 6200 yr B.P., occurrence of marine elements in a freshwater, palustral community; (c) until 5500 years B.P., replacement of the sweetwater community by halophyte; (d) until 4800 years B.P., increasing marine influence; (e) until 4120 years B.P., maximum marine influence in the stream valley.

4. Pardiñas (2001) described the site "Camping Americano", to the southwest of Monte Hermoso locality, where they found fish, amphibians, birds, marsupials, armadillos and rodents; faunal elements that suggest xeric climatic conditions in interdune littoral lagoons. A radiocarbon dating provided an age of 8.9 ka years B.P.

   Aramayo et al. (2005) studied the coastal evolution between Monte Hermoso and Pehuen-Co localities, distinguishing for Holocene a succession of environments: (a) continental, formed by sandy silt lenticular bodies; (b) beaches and coastal lagoons with caves of crab Chasmagnathus (which does not reach the zone at present), carbonized stems of Frankenia juniperoides, which currently inhabits from Bahía Blanca to the south (Correa 1966) and Spartina colonization.

5 Iriondo's studies (1991) pointed out the occurrence of significant climatic changes for Mesopotamia and a "Paranaense" strip of Santa Fe Province:

(a) Late Pleistocene climate, drier and colder with intense eolian action and savanna landscape in Misiones Province;

(b) Early and Middle Holocene, wetter with higher temperatures than the present ones. Late Holocene, drier subtropical climate, caused by an anticyclone in the Pampean region and then changed to the current climate.

6. In the Montenegro locality, southern Brazil, the fauna findings denoted subtropical environments and the presence of forests (Oliveira 1999).

7. Meyer et al. (2000) identified four palynologic zones for the Holocene in core samples of the Itapeva lake in Rio Grande do Sul: (1) swamp and continental water environment; (2) penetration of marine waters (5100 years B.P.); (3) saline swamp during marine regression; domain of continental waters.

The Platan ichnological record is varied; it is represented by two ichnofacies:

(a) Scoyenia ichnofacies.

1. Human footprints in environments of littoral lagoons, in Monte Hermoso, 8000 years ago.

2. Chasmagnathus crab habitats in Pehuen-Có. Nowadays, the genus only reaches to the Mar Chiquita lagoon.

(b) Coprinispha ichnofacies

1. Upper levels of Cueva Tixi, in the Tandilia hills

3. Mima type mounds, in the Necochea region.

4. Coprinisphaera ichnofacies represented by Celliformidae family, in the Paraná gullies.

Paula Couto (1982) observed that Hylea as biotic formation is relatively recent, regarding to its origin. Pleistocene fauna found in the rivers Napo, Ucayali and Jurua in the Brazilian state of Acre (Rancy 1991) showed that the Amazon basin was fragmented regarding its rain forest vegetation, resulting in savanna zones with rain forests in galleries of the existing rivers.

Tonni and Cione (1997) agreed, concerning the installation of present conditions prevailing in the Pampean plain.

References

Aceñolaza F, Aceñolaza G (2004) Trazas fósiles en unidades estratigráficas del Neógeno de Entre Ríos. In: Temas de la biodiversidad del Litoral fluvial argentino, INSUGEO, Miscelánea, vol 12. San Miguel de Tucumán, pp 19–24

Alberdi M, Prado J (1995a) Los Mastodontes de América del Sur. In: Alberdi M, Leone G, Tonni E (eds) Evolución biológica y climática de la región pampeana durante los últimos cinco millones de años. Monografía N° 12. Museo Nacional de Ciencias Naturales, Madrid. pp 279–291

Alberdi M, Prado J (1995b) Los équidos de América del Sur. In: Alberdi M, Leone G, Tonni E (eds) Evolución biológica y climática de la región pampeana durante los últimos cinco millones de años. Monografía N° 12. Museo Nacional de Ciencias Naturales. Madrid. pp 295–307

Albino A, Carlini A (2007) Boa constrictor Linnaeus (Serpentes, Boidae) en el Pleistoceno tardío del Arroyo Toropí (Corrientes, Argentina). Ameghiniana, Suplemento Resúmenes 44(4):77R. Buenos Aires

Albino A, Montalvo C (2006) Snakes from the Cerro Azul Formation (Upper Miocene), Central Argentina, with a review of fossil viperids from South America. J Paleontol 26(3):581–587

Andreis R (1965) Petrografía y paleocorrientes de la Formación Río Negro. Revista del Museo de La Plata, n.s. Geología 36. La Plata

Antón D (1975) Evolución geomorfológica del norte del Uruguay. Dirección de suelos y Fertilizantes. Ministerio de Agricultura y Pesca. Montevideo, unpublished internal report, pp 1–22

Aramayo S, Gutiérrez Téllez B, Schillizzi RA (2005) Sedimentologic and paleontologic study of the southeast coast of Buenos Aires province, Argentina: a late pleistocene holocene paleoenvironmental reconstruction. J S Am Earth Sci 20:65–71

Araujo RL (1970) Termites of the Neotropical region. In: Krishna, Weesner FM (eds) Biology of termites, vol 2. Academic Press. New York, pp 527–571

Báez A (1986) El registro terciario de los anuros en territorio argentino. Una revaluación. IV Congreso Argentino de Paleontología y Bioestratigrafía, Actas vol 2. Mendoza, pp 107–118

Barreda V, Anzótegui LM, Prieto AR, Aceñolaza P, Bianchi MM, Borromei AM, Brea M, Caccavari M, Cuadrado GA, Garralla S, Grill SG, Guerstein GR, Lutz AI, Mancini MV, Mautino R, Ottone EG, Quattrocchio ME, Romero EJ, Zamaloa MC, Zucol A (2007) Diversificación y cambios de las Angiospermas durante el Neógeno en Argentina, In: Archangelsky S, Sánchez T, Tonni E (eds) Asociación Paleontológica Argentina, Publicación Especial 11, Ameghiniana, 50° aniversario, pp 173–191

Bombín M (1976) Modelo paleoecológico evolutivo para o Neoquaternario da regiao da campanha-oeste de Río Grande do Sul (Brasil), a Formaçao Touro Passo, seu conteúdo fossilifero e a pedogénese pós deposicional. Comunicacoes Do Mus. Cs. da PUCRGS, vol 15. Porto Alegre, pp 1–90

Bond M (1986) Los carnívoros terrestres fósiles de Argentina: Resumen de su historia. In: IV Congreso Argentino de Paleontología y Bioestratigrafía. Actas vol 2. Buenos Aires, pp 173–185

Bond M (1999) Quaternary native ungulates of Southern South America. A synthesis. In: Rabassa J, Salemme M (eds) Quaternary vertebrate paleontology in South America. Special volume of Quaternary of South America and Antarctic Peninsula, vol 12. A.A. Balkema Publishers, Rotterdam, pp 177–205

Bond M, Perea D, Ubilla M, Tauber A (2001) Neolicaphrium recens Frenguelli 1921, the only surviving Protherotheriidae (Litopterna, Mammalia) into the South American Pleistocene. Palaeovertebrata 30(1–2):37–50

Bondesio P, Laza J, Scillato Yané G, Tonni E, Vucetich G (1980) Estado actual del conocimiento de los vertebrados de la Formación Arroyo Chasicó (Plioceno temprano) de la provincia de Buenos Aires. In: II Congreso Argentino de Paleontología, Bioestratigrafía and 1°Congreso Latinoamericano de Paleontología. Actas, vol III. Buenos Aires, pp 101–127

Bown T, Kraus M (1987) Integration of channel and floodplain suites. I Developmental sequence and lateral relations of alluvial paleosols. J Sediment Petrol 57:587–601

Brea M, Zucol A (2007) New fossil record from Uruguay basin related to El Palmar Formation floristic composition. Ameghiniana 44(4) Suplemento Resúmenes:78R. Buenos Aires

Bucher E (1980) Ecología de la fauna chaqueña. Una revisión. Ecosur 7(14):111–159

Buchmann FS, Pereira Lopes R, Caron F (2009) Icnofosseis (paleotocas e crotovinas) atribuidos a mamíferos extintos no sudeste e sul do Brasil. Rev Brasileira de Paleontol 12:247–256

Cabrera A, Willink A (1973) Biogeografía de América Latina. Monografía N° 13. Serie Biología, OEA, Washington DC

Carlini A, Scillato-Yané G (1999) Evolution of Quaternary Xenarthrans (Mammalia) of Argentina. In: Rabassa J, Salemme M (eds) Quaternary vertebrate palaeontology in South America. Special volume of Quaternary of South America and Antarctic Peninsula, vol 12. A.A. Balkema Publishers, Rotterdam, pp 149–175

Carlini A, Zurita A, Gasparini G, Noriega J (2004) Los mamíferos del Pleistoceno de la Mesopotamia argentina y su relación con los del Centro-Norte de la Argentina, Paraguay, Sur de Bolivia, y los del Sur de Brasil y Oeste de Uruguay: Paleobiogeografía y Paleoambientes. INSUGEO, Miscelánea, vol 12. San Miguel de Tucumán, pp 83–90

Casamiquela R (1974) El bipedismo de los megaterioideos. Estudio de pisadas fósiles en la Formación Río Negro típica. Ameghiniana 11(3):249–282. Buenos Aires

Cavallotto JL (1995) Evolución de la topografía del sustrato del Holoceno del "Río de La Plata" 4° Jornadas Geológicas y Geofísicas Bonaerenses, Actas 1:223–229. La Plata

Cenizo M, Montalvo C (2006) Nuevos registros de aves para la Formación Cerro Azul, Mioceno Tardío, provincia de La Pampa, Argentina. Ameghiniana 43(4): Suplemento Resúmenes: 29. Buenos Aires

Chani S (1976) Relaciones de un nuevo Teiidae (Lacertilia) fósil del Plioceno superior, Callopistes bicuspidatus n.sp. In: Instituto Miguel Lillo, Universidad Nacional de Tucumán, Publicación Especial. San Miguel de Tucumán, pp 133–153

Cione A, Báez A (2007) Peces continentales y anfibios cenozoicos de Argentina: los últimos cincuenta años. In: Ameghiniana, Publicación Especial 11 (50° aniversario). Buenos Aires, pp 195–220

Cione A, López Arbarello A (1995) Los peces de agua dulce fósiles del área pampeana (131–142). In: Alberdi M, Leone G, Tonni E (eds) Evolución biológica y climática de la región pampeana durante los últimos cinco millones de años. Monografía N° 12, Museo Nacional de Ciencias Naturales, Madrid. p 423

Cione A, Tonni E (1995b) Bioestratigrafía y cronología del Cenozoico superior de la Región pampeana (47–74). In: Alberdi M, Leone G, Tonni E (eds) Evolución biológica y climática de la región pampeana durante los últimos cinco millones de años. Monografía N° 12, Museo Nacional de Ciencias Naturales, Madrid. p 423

Cione A, Tonni E (1999) Biostratigraphy and chronological scale of uppermost Cenozoic in the Pampean area, Argentina (23–52). In: Rabassa J, Salemme M (eds) Quaternary vertebrate paleontology in South America. Special volume of Quaternary of South America and Antarctic Peninsula, vol 12, A.A. Balkema Publishers, Rotterdam, p 310

Cione A, Tonni E (2001) Correlation of Pliocene to Holocene Southern South American and European vertebrate-bearing units. In: Rook L, Torre D (eds) Neogene and quaternary continental stratigraphy and mammal evolution, vol 40 (2). Bolletino della Societá Paleontologica Italiana, pp 167–173

Cione A, Tonni E (2005) Bioestratigrafía basada en mamíferos del Cenozoico superior de la provincia de Buenos Aires, Argentina. In: XVI Congreso Geológico Argentino, Relatorio. La Plata, pp 183–200

Cione A, Azpelicueta M, Bond M, Carlini A, Casciotta J, Cozzuol M, de la Fuente M, Gasparini Z, Goin F, Noriega J, Scillato Yané G, Soibelzon L, Tonni E, Verzi D, Vucetich M (2000) Miocene vertebrates from Entre Ríos, eastern Argentina. In: Aceñolaza G, Herbst R (eds) El Neógeno en Argentina. INSUGEO Serie Correlación Geológica, vol 14(1–2). San Miguel de Tucumán, pp 191–237

Cione A, Tonni E, San Cristóbal J (2002) A Middle-Pleistocene marine transgression in Central-eastern Argentina. CRP 19:16–18

Cione A, Tonni E, Soibelzon L (2003) The broken zig-zag: late cenozoic large mammal and turtle extinction in South America. In: Revista del Museo Argentino de Ciencias Naturales "Bernardino Rivadavia", vol 5. Buenos Aires, pp 1–19

Cione A, Tonni E, Bargo S, Bond M, Candela A, Carlini A, Deschamps C, Dozo T, Esteban G, Goin F, Montalvo C, Nasif J, Ortiz Jaureguizar C, Pascual R, Prado J, Reguero M, Scillato Yané G, Soibelzon E, Verzi D, Vieytes E, Vizcaino S, Vucetich M (2007) Mamíferos continentales del Mioceno tardío a la actualidad en la Argentina: cincuenta años de estudio. In: Ameghiniana, Publicación Especial 11 (50° aniversario). Buenos Aires, pp 257–278

Cione AL, Gasparini GM, Soibelzon E, Soibelzon L, Tonni EP (eds) (2015) The great American biotic interchange. A South American perspective. Springer Briefs in Earth system sciences. Springer, The Netherlands, p 97

Constantino R (1999) Chave ilustrada para identificaçao dos géneros de cupins (Insecta, Isoptera) que ocorren no Brasil. Papeis Avulsos de Zoología 40(25):387–448. Sao Paulo, Brasil

Correa M (1966) Las Frankeniaceae argentinas. Darwiniana 14(1):68–94. Buenos Aires

Croizat L (1976) Biogeografía analítica y sintética ("panbiogeografía") de las Américas Biblioteca de la Academia de Ciencias Físicas. Matemáticas y Nat 15–16:455–890. Caracas, Venezuela

Darwin C (1844) La teoría de la evolución de las especies. Ensayo. In: De Agostini (ed) Darwin, vida, pensamiento y obra. 2007, Spain

De La Fuente M (1999) 8. A review of the Pleistocene reptiles of Argentina: taxonomic and palaeoenvironmental considerations (109–136). In: Rabassa J, Salemme M (eds) Quaternary vertebrate palaeontology in South America. Special Volume of Quaternary of South America and Antarctic Peninsula N°12, A.A. Balkema Publishers, Rotterdam, pp. 109–136

de Paula Couto C (1982) Fossil Mammals from the Cenozoic of Acre, Brazil. Part 5, Notoungulata Nesodontinae, Haplodontheriinae and Litopterna, Pyrotheria, and Astrapotheria. Iheringia ser Geol 7:5–43

Deschamps C (1995) Late Cenozoic mammal bio-chronostratigraphy in southwestern Buenos Aires Province, Argentina. Ameghiniana 42(4):733–750. Buenos Aires

Donadio O (1984) Los lacertilios fósiles de la provincia de Cordoba (Sauria-Teiidae) y sus implicaciones paleoambientales. In: III Congreso Argentino de Paleontología, Bioestratigrafía, Actas. Buenos Aires, pp 217–223

Edmond A (1985) The armor of fossil giant armadillos (Pampatheriidae, Xenarthra, Mammalia). In: Texas memorial museum, university of Texas and Austin, Pearce Sellards Series, vol 40. Lubbock, pp 1–20

Eiten G (1972) The cerrado vegetation of Brasil. Bot Rev 38(2):201–341

Eiten G (1997) Uso do termo Savana. Portugaliae Acta Biol Ser B Sist 17:1–262

Emerson A (1952) The neotropical genera Procornitermes and Cornitermes (Isoptera Termitidae). Bull Am Mus N.H. 99:475–540

Erra G, Osterrieth M, Zurita A, Lutz A, Laffont E, Coronel J, Francia A (2016) Primer registro de termiteros fósiles para el Pleistoceno tardío de la región Mesopotámica (Argentina): implicancias paleoambientales. Acta Biol Colomb 21(1):63–72. Bogotá

Fasano J, Isla F, Schnack E (1984) Significado paleoambiental de los depósitos del Pleistoceno tardío de Camet Norte (Partido de Mar Chiquita, provincia de Buenos Aires. Rev Asoc Geol Argentina 39(3–4):244–261. Buenos Aires

Ferreira Brandao C, Vanzolini P (1985) Notes on incubatory inquilinism between Squamata (Reptilia) and the Neotropical fungus-growing ant genus Acromyrmex (Hymenoptera: Formicidae). Papéis Avulsos de Zoologia 36(3):31–36

Ferrero B (2007) Los mastodontes (Mammalia, Gomphotheriidae) en el Lujanense de la Provincia de Entre Ríos. Consideraciones geográficas y paleoecológicas. In: XXIII Jornadas Argentinas de Paleontología de Vertebrados. Resúmenes: 13. Trelew, Argentina

Fidalgo F (1993) Provincia de Buenos Aires. Continental (23–38). In: Iriondo M (ed) El Holoceno en la Argentina, CADINCUA. Paraná

Fidalgo F, Colado U, De Francesco F (1973a) Sobre las ingresiones marinas cuaternarias en los partidos de Castelli, Chascomús y Magdalena (provincia de Buenos Aires). In: V Congreso Geológico Argentino. Actas III: Buenos Aires, pp 227–240

Fidalgo F, Meo Guzmán L, Politis G, Salemme M, Tonni E (1986) Investigaciones arqueológicas en el Sitio 2 de Arroyo Seco (Partido de Tres Arroyos—Provincia de Buenos Aires—República Argentina). In: Bryan AL (ed) New evidence for the pleistocene people of the Americas. Center for the study of Early Man. University of Maine at Orono, Orono, pp 221–269

Fidalgo F, Tonni E, Porro N, Laza J (1987) Geología del área de la Laguna Chasicó (Partido de Villarino, provincia de Buenos Aires) y aspectos bioestratigráficos relacionados. Rev Asoc Geol Argentina 42(3–4):407–416. Buenos Aires

Frenguelli J (1921) Los terrenos de la costa Atlántica en los alrededores de Miramar (Provincia de Buenos Aires) y sus correlaciones. Acad Nac de Cienc de Córdoba 24:325–485. Córdoba

Frenguelli J (1928) Observaciones geológicas en la región costera sur de la Provincia de Buenos Aires. Universidad Nacional del Litoral. Anales Fac de Cienc de la Educación 3:101–130. Santa Fe

Frenguelli J (1938a) Bolas de escarabeidos y nidos de véspidos fósiles. Physis 12:348–352. Buenos Aires

Gasparini Z, Báez A (1975) Aportes al conocimiento de la herpetofauna terciaria de la Argentina. I Congreso Argentino de Paleontología y Bioestratigrafía. Actas 2:377–415. San Miguel de Tucumán

Gasparini Z, de la Fuente M, Donadío O (1986) Los reptiles cenozoicos de la Argentina: implicancias paleoambientales y evolución biogeográfica. IV Congreso Argentino de Paleontol y Bioestratigrafía. Actas 2:119–130. Buenos Aires

Genise J (1989) Las cuevas con Actenomys (Rodentia, Octodontidae) de la Formación Chapadmalal (Plioceno Superior) de Mar del Plata y Miramar (provincia de Buenos Aires). Ameghiniana 26 (1–2):33–4. Buenos Aires

Goin F, Montalvo C, Visconti G (2000) Los marsupiales (Mammalia) del Mioceno Superior de la Formación Cerro Azul (Provincia de La Pampa, Argentina). Estud Geol 56:101–126

González Bonorino F (1965) Mineralogía de las fracciones arcilla y limo del Pampeano en el área de la ciudad de Buenos Aires y su significado estratigráfico y sedimentológico. Rev de la Asoc Geol Argentina 20:67–148. Buenos Aires

Haffer J (1969) Speciation in Amazonian forest birds. Science 165:131–137

Halffter G, Edmonds WD (1982) The nesting behavior of dung beetles (Scarabaeinae): an ecological and Evolutive approach. Instituto de Ecología, México D.F, p 176

Herbst R, Brea M, Crisafulli A, Gnaedinger S, Lutz A, Martínez C (2007) La paleoxilología en la Argentina. Historia y desarrollo. Ameghiniana, Publicación especial 50° aniversario 11:57–71. Buenos Aires

Iriondo M (1980) El Cuaternario de Entre Ríos. Rev de la Asoc de Cienc Nat del Litoral 11:125–141. Santa Fe

Iriondo M (1991) El Holoceno en el Litoral. Comun del Museo Provincial de Cienc Nat "Florentino Ameghino" n.s. 3(1):1–39. Santa Fe

Iriondo M (2007) El Chaco santafesino. Neógeno y Geomorfología. Museo Provincial de Cienc Naturales "Florentino Ameghino", Comun 13(1):1–39. Santa Fe

Iriondo M, Krohling D (1996) Los sedimentos eólicos del noreste de la llanura pampeana (Cuaternario superior). In: XIII Congreso Geológico Argentino and III Congreso de exploración de hidrocarburos. Actas, vol 4. Buenos Aires, pp 27–48

Iriondo M, Krohling D (2001) A neoformed kaolinitic mineral in the Upper Pleistocene of NE Argentina. Int Clay Conf Abstr 12:6

Isla F, Cortizo L, Schnack E (1996) Pleistocene and Holocene beaches and estuaries along the southern barrier of Buenos Aires, Argentina. Quat Sci Rev 13:833–841

Kraglievich J (1952) El perfil geológico de Chapadmalal y Miramar, provincia de Buenos Aires. Rev Museo Municipal Cienc Nat Tradiciones, Mar del Plata 1(1):8–73

Kraus M (1988) Nodular remains of early Tertiary forest, Bighorn Basin, Wyoming. J Sediment Petrol 58:888–893

Kusnezov N (1951) El género Pheidole en la Argentina. (Hymenoptera, Formicidae). Acta Zool Lilloana 12:5–88. San Miguel de Tucumán

Kusnezov N (1963) Zoogeografía de las hormigas en Sudamérica. Acta Zool Lilloana 19:25–186. San Miguel de Tucumán

Laza J (1995) Signos de actividad de insectos. In: Alberdi M, Leone G, Tonni E (eds) "Evolución biológica y climática de la región pampeana durante los últimos cinco millones de años". Museo Nac de Cienc Nat Madrid. Monografía N° 12. pp 341–361

Laza J (2001) Nidos de Scarabaeinae. Significación paleoclimática y cronológica. In: Mazzanti D, Quintana C (eds) Cueva Tixi: cazadores y recolectores de las sierras de Tandil Oriental I. Geología, Paleontología y Zooarqueología. p 119

Linares E, Llambías E, Latorre C (1980) Geología de la provincia de La Pampa, República Argentina y geocronología de sus rocas metamórficas y eruptivas. Rev de la Asoc Geol Argentina 35:87–146. Buenos Aires

Malagnino E (1989) Paleoformas de origen eólico y sus relaciones con los modelos de inundación de la provincia de Buenos Aires. In: IV Simposio latinoamericano de percepción remota and IX Reunión Plenaria Selper. Tomo II: San Carlos de Bariloche, pp 611–620

Mares MA, Rosenzweig ML (1978) Granivory in north and south American deserts: rodents, birds, and ants. Ecology 59(2):235–241

Martínez A (1959) Catálogo de los Scarabaeidae argentinos. Rev del Museo Argentino de Cienc Nat "Bernardino Rivadavia", Cs. Zool. V(1):1–126. Buenos Aires

Melchor R, Visconti G, Montalvo C (2000) Late Miocene calcic vertisols from central La Pampa, Argentina. II Congreso Latinoam de Sedimentología y VII Reunión Argent de Sedimentología, Actas. Buenos Aires, pp 119–120

Menegaz A, Ortiz Jaureguizar E (1995) Los Artiodáctilos. In: Alberdi M, Leone G, Tonni E (eds) Evolución biológica y climática de la región pampeana durante los últimos cinco millones de años. Monografía N°12, Museo Nacional de Ciencias Naturales, Madrid, pp 309–338

Meyer K, Reichhart A, Ashraf M, Marques-Toigo M, Mosbrugger V (2000) Holocene paleoenvironmental evolution or Northern Coastal Plain of Río Grande do Sul State, Brasil: palynological and geochemical records. Ameghiniana 37 (4) suplemento:57R

Montalvo C, Verzi D, Vucetich M, Visconti G (1998) Nuevos eumysopinae (Rodentia, Echymiidae) de la Formación Cerro Azul. (Mioceno tardío) de La Pampa, Argentina. V Jorn Geol y Geofísicas bonaerenses, Actas I:57–63. La Plata

Morrone J (2000) What is the chacoan subregion? Neotropica 46:51–68

Morrone J (2001) Biogeografía de América Latina y el Caribe. SEA-CYTED, Mexico, Spain, p 144

Nabel P, Cione A, Tonni E (2000) Environmental changes in the Pampean area of Argentina at the Matuyama-Brunhes (C1r–C1n) chrons boundary. Palaeogeogr Palaeoclimatol Palaeoecol 162:403–412

Netto R, Tognoli F, de Gibert J, de Oliveira M (2007) Paleosol evolution in Nearshore fluviatile Pleistocene deposits of the Chui Formation, South of Brazil. In: Quinta Reunión Argent de Icnología y Tercera Reunión de Icnología del Mercosur. Resúmenes: 55. Ushuaia, Tierra del Fuego. Argentina

Noriega J, Areta J (2005) First record of Sarcoramphus Dumeril 1806 (Ciconiformes: Vulturidae) from the Pleistocene of Buenos Aires province, Argentina. J S Am Earth Sci 20:73–79

Noriega J, Carlini A, Tonni E (2004) Vertebrados del Pleistoceno tardío de la cuenca del Arroyo Ensenada, Departamento Diamante, Provincia de Entre Ríos. Temas de la biodiversidad del Litoral fluvial argentino. INSUGEO, Miscelanea 12:71–76

Oliveira E (1992) Mamíferos fósseis do Quaternário do Estado do Río Grande do Sul, Brasil. Curso de Pós-Graduacao em Geociencias, Universidade Federal do Río Grande do Sul, Porto Alegre. Unpublished

Oliveira E (1999) 5. Quaternary vertebrates and climates of southern Brasil. In: Rabassa J, Salemme M (eds) J. Quaternary vertebrate paleontology in South America. Special volume of Quaternary of South America and Antarctic peninsula, 12. pp 61–73

Pardiñas U (1999) 13. Fossil murids: taxonomy, palaeoecology and palaeoenvironments. In: Rabassa J, Salemme M (eds) Quaternary vertebrate palaeontology in South America, Special volume of Quaternary of South America and Antartic Peninsula, 12. pp 225–254

Pardiñas U (2001) Condiciones áridas durante el Holoceno Temprano en el sudoeste de la provincia de Buenos Aires (Argentina). Vertebrados y tafonomía. Ameghiniana 38(3):227–236

Pardiñas U, Lezcano M (1995) Cricétidos (mammalia, rodentia) del Pleistoceno tardío del nordeste de la provincia de Buenos Aires (Argentina). Aspectos sistemáticos y paleoambientales. Ameghiniana 32(3):249–265

Pardiñas U, Gelfo J, San Cristóbal J, Cione A, Tonni E (1996) Una asociación de organismos marinos y continentales en el Pleistoceno superior en el sur de la provincia de Buenos Aires, Argentina. In: XIII Congreso Geológico Argentino and III Congreso de Exploración de Hidrocarburos, Actas. Buenos Aires, pp 95–111

Pardiñas U, Cione A, San Cristóbal J, Verzi D, Tonni E (2004) A new last interglacial continental vertebrate assemblage in Central Eastern Argentina. Curr Res Pleistocene 21:111–112

Parker G, Paterlini CM, Violante RA (1994) Edad y génesis del Río de La Plata. Rev de la Asoc Geol Argent 49:11–18

Pascual R (1961) Un nuevo cardiomyinae (Rodentia, Caviidae) de la formación arroyo chasicó (Plioceno inferior) de la provincia de buenos aires. Ameghiniana 2(4):57–71

Pascual R, Bond M (1986) Evolución de los marsupiales cenozoicos de Argentina. IV Congreso Argentino de Paleontología y Bioestratigrafía. Actas 2:143–150

Pascual R, Bondesio P (1982) Un roedor Cardiatheriinae (Hydrochoeriidae) de la Edad Huayquerense (Plioceno tardío) de La Pampa. Sumario de los ambientes terrestres de la Argentina durante el Mioceno. Ameghiniana, 19(1–2):19–35

Pereira F, Martínez A (1956) Algunas notas sinonímicas en Phanaeini. Rev Bras de Entomología 5:229–240

Peters J, Donoso Barros R (1970) Catalogue of the Neotropical Squamata. Parte II. Lizards and Amphisbenians. Smithsonian Inst. U.S. Nat Mus Bull 297:1–293

Pillans B, Naish T (2004) Defining the quaternary. Quatern Sci Rev 23:2271–2282

Politis G (1984) Climatic variations during historical times in eastern buenos aires pampas, argentina. Quat S Am Antartic peninsula 2(9):133–162

Politis G (1993) Las pisadas humanas de Monte Hermoso dentro del contexto de la arqueología pampeana. Primera reunión argentina de icnología. La Pampa, Argentina. Resúmenes y conferencias: 25

Pomi L, Tonni E (2010) Marcas de insectos sobre huesos del Pleistoceno tardío de la Argentina. X Congreso Argentino de Paleontología y Bioestratigrafía y VII Congreso Latinoamericano de Paleontología. La Plata. Universidad Nacional de La Plata, Museo de La Plata. Resúmenes: (n°342):201

Preciozzi F, Spoturno J, Heinzen W, Rossi P (1985) Memoria explicativa de la Carta Geológica del Uruguay a la escala 1: 500.000. DINAMIGE, Montevideo, Uruguay, pp 1–90

Prieto A (1996) Late quaternary vegetational and climatic changes in the Pampa grassland of Argentina. Quatern Res 45:73–88

Prieto A (2000) Vegetational history of the Late glacial-Holocene transition in the Grasslands of eastern Argentina. Palaeogeogr Palaeoclimatol Palaeoecol 157:167–188

Quattrochio M, Borromei A, Grill S (1995) Cambios vegetacionales y fluctuaciones paleoclimáticas durante el Pleistoceno tardío—Holoceno en el sudoeste de la Provincia de Buenos Aires (Argentina). VI Congreso Argentino de Paleontología y Bioestratigrafía, Actas, pp 221–229

Rabassa J (2008) Late cenozoic glaciations in patagonia and tierra del Fuego. In: Rabassa J (eds) Late cenozoic of patagonia and tierra del fuego, Developments in Quaternary Science. Elsevier, p 11

Rancy A (1991) Pleistocene mammals and paleoecology of the Western Amazon. Unpublished Ph.D. dissertation, University of Florida, Gainesville

Redford K (1984) The termitaria of Cornitermes cumulans (Isoptera: Termitidae) and their role in determining a potencial keystone species. Biotropica 16(2):112–119

Redford K (1986) The role of gallery forests in the zoogeography of the Cerrado's non-volant mammalian fauna. Biotropica 18(2):126–135

Retallack G (2001) Cenozoic expansion of grassland and climatic cooling. J Geol 109:407–426

Ringuelet R (1961) Rasgos fundamentales de la zoogeografía de la Argentina. Physis 22
Ringuelet R (1975) Zoogeografía y ecología de los peces de aguas continentales de Argentina y consideraciones sobre las áreas ictiológicas de América del Sur. Ecosur 3:1–122
Ringuelet R (1978) Dinamismo histórico de la fauna brasílica en la Argentina. Ameghiniana 15(1–2):255–262
Romero E, Fernández C (1981) Palinología de paleosuelos del Cuaternario de los alrededores de Lobería (Provincia de Buenos Aires, República Argentina). Ameghiniana 18(3–4):273–285
Rossi V, Osterrieth M, Martínez G (2001) Icnología de la secuencia estratigráfica Mar Chiquita. (Cuaternario tardío), Buenos Aires. In: IV Reunión Argentina de Icnología y Segunda Reunión de Icnología del Mercosur. Resúmenes: 68. Tucumán
Sánchez MV (2009) Trazas fósiles de coleópteros coprófagos del Cenozoico de la Patagonia Central. Significado evolutivo y paleoambiental. Unpublished doctoral thesis. Facultad de Ciencias Exactas y Naturales, Universidad Nacional de Buenos Aires
Sánchez MV, Genise J (2009) Cleptoparasitismo and detritivory in dung beetle fossil brood balls from Patagonia, Argentina. Palaeontology 52(4):837–848
Sarmiento G (1992) Adaptative strategies of perennial grasses in South American Savannas. J Veg Sci 3:325–336
Schabitz F (2003) Estudios polínicos del Cuaternario en las regiones áridas del sur de Argentina. Rev del Museo Argentino de Cienc Nat n.s. 5 (2):291–299
Schnack E, Isla F, De Francesco F, Fucks E (2005) Estratigrafía del Cuaternario marino tardío en la provincia de Buenos Aires. XVI Congreso Geológico Argentino, Relatorio X: 159–182. Buenos aires
Scillato Yané G (1986) Los Xenarthra fósiles de Argentina (Mammalia, Edentata). In: IV Congreso Argentino de Paleontología y Bioestratigrafía, Actas vol 2. Mendoza, pp 151–155
Silvestri F (1903) Contribuzione alla conoscenza dei Termiti e Termitofili dell America Meridionale. REDIA 1:1–234
Simpson G (1970) The Argyrolagidae, extinct South American marsupials. Bull Mus Comp Zool Harvard Univ 139:1–86
Soibelzon L (2009) Científicos platenses revelan datos de osos ya extinguidos. Agencia Noticiosa Cyta, Instituto Leloir, Diario El Día, 8–8-09, La Plata
Stutz S, Prieto A, Isla F (1999) Cambios de la vegetación durante el Holoceno en el SE de la provincia de Buenos Aires: análisis polínico del arroyo La Ballenera. Asociación Paleontológica Argentina. Publicación Especial 6. X Simposio Argentino de Paleobotánica y Palinología, pp 65–69
Tambussi C (1995) Las aves. In: Alberdi M, Leone G, Tonni E (eds) Evolución biológica y climática de la región pampeana durante los últimos cinco millones de años. Monografías N°12. Museo Nacional de Ciencias Naturales, España. pp 145–161
Teruggi M, Imbellone P (1987) Paleosuelos loéssicos superpuestos en el Pleistoceno Superior-Holoceno de la región de La Plata, Provincia de Buenos Aires vol 5. Ciencias del Suelo, pp 175–188
Teruggi M, Imbellone P (1988) Paleosuelos de la región pampeana. Segunda jornada de suelos de la región pampeana. La Plata. Actas, pp 40–66
Tonni E (1980) The present state of knowledge of the Cenozoic birds of Argentina. Cont Sci Natur Hist Mus Los Angeles County 330:105–114
Tonni E (1987) Stegomastodon platensis y la antiguedad de la Formación El Palmar, en el Departamento Colón, Entre Ríos. Ameghiniana 24:323–324
Tonni E (1992a) Tapirus Brisson, (1762) (Mammalia, Perissodactyla) en el Lujanense (Pleistoceno superior-Holoceno inferior) de la provincia de Entre Ríos, República Argentina. Ameghiniana 29(1):3–8
Tonni E (1992b) Mamíferos y clima del Holoceno en la provincia de Buenos Aires. In: Iriondo M (ed) CADINQUA, vol 1. El Holoceno en la Argentina, pp 64–88

Tonni E (2009) Las unidades portadoras de vertebrados del Cuaternario de las regiones mesopotámica y pampeana oriental de la Argentina. In: Ensayo de correlación. Quaternario do Río Grande do Sul. Monografía da Sociedade Brasileira de Paleontología. pp 57–66

Tonni E, Cione A (1984) A thanatocenosis of continental and marine vertebrates in the Las Escobas Fm. (Holocene) of Northeastern Buenos Aires Province, Argentina. Quat S Am Antarctic Peninsula, 2:93–113. Rotterdam: A.A. Balkema

Tonni E, Cione A (1995) 18. Los mamíferos como indicadores de cambios climáticos en el Cuaternario de la región pampeana de la Argentina. In: Argollo, Mourguiart (eds) La Paz, Bolivia. Cambios Cuaternarios en América del Sur. pp 319–326

Tonni E, Cione A (1997) Did the Argentine Pampean ecosysten exist in the Pleistocene? Curr Res Pleistocene 14:131–133

Tonni E, Fidalgo F (1978) Consideraciones sobre los cambios climáticos durante el Pleistoceno tardío-reciente en la provincia de Buenos Aires. Aspectos ecológicos y zoogeográficos relacionados. Ameghiniana 15(1–2):235–253

Tonni E, Fidalgo F (1982) Geología y paleontología de los sedimentos del Pleistoceno en el área de Punta Hermengo (Miramar, Provincia de Buenos Aires, República Argentina). Ameghiniana 19(1–2):79–108

Tonni E, Laza J (1980a) Las aves de la Fauna Local Paso de Otero (Pleistoceno tardío de la provincia de Buenos Aires). Su significación ecológica, climática y zoogeográfica. Ameghiniana 17(4):313–322

Tonni E, Noriega J (1996) Una nueva especie de Nandayus Bonaparte, 1854 (Aves: Psittaciformes) del Plioceno tardío de Argentina. Rev Chil de Historia Nat 69:97–104

Tonni E, Politis G (1982) Un gran cánido del Holoceno de la provincia de Buenos Aires y el registro prehispánico de Canis (Canis) familiares en las áreas Pampeana y Patagónica. Ameghiniana 18(3–4):251–265

Tonni E, Tambussi C (1986) Las aves del Cenozoico de la República Argentina. IV Congreso Argentino de Paleontología y Bioestratigrafía. Actas 2:131–142

Tonni E, Verzi D, Bargo M, Pardiñas J (1993). Micromammals in owl pellets from the lower-Middle pleistocene in buenos aires province, Argentina. X Jornadas de Paleontología de Vertebrados. Resúmenes. Ameghiniana 30(3):342

Tonni E, Verzi D, Bargo M, Scillato Yané G, Pardiñas U (1995) Bioestratigrafía del Cenozoico superior continental en las barrancas costeras de Necochea y Miramar, Provincia de Buenos Aires, República Argentina. IV Jornadas Geológicas y Geofísicas Bonaerenses. Actas 1, pp 63–71

Tonni E, Pardiñas U, Verzi D, Noriega J, Scaglia O, Dondas A (1998) Microvertebrados pleistocénicos del sudeste de la provincia de Buenos Aires (Argentina): bioestratigrafía y paleoambientes. V Jornadas geológicas y geofísicas bonaerenses. Actas I:73–83

Tonni E, Nabel P, Cione A, Etchichury M, Tófalo R, Scillato Yané G, San Cristóbal J, Carlini A, Vargas D (1999a) The ensenada and buenos aires formations (Pleistocene) in the quarry near La Plata, Argentina. J S Am Earth Sci 12:273

Tonni E, Cione A, Figini A (1999) Predominance of arid climates indicated by mammals in the pampas of Argentina during the late Pleistocene and Holocene. Palaeogeogr Palaeoclimatol Palaeoecol 147:257–281

Tonni E, Cione A, Figini A (2001) Chronology of Holocene pedogenetic events in the Pampean area of Argentina. Curr Res Pleistocene 18:124–127

Tonni E, Huarte R, Carbonari J, Figini A (2003) New radiocarbon chronology for the Guerrero Member of the Luján Formation (Buenos Aires, Argentina): paleoclimatic significance. Quatern Int 109–110:45–48

Tonni E, Carlini A, Zurita A, Frechem M, Gasparini G, Budziak D, Kruck W (2005) Cronología y Bioestratigrafía de las unidades del Pleistoceno aflorantes en el Arroyo Toropí, Provincia de Corrientes, Argentina. XIX Congreso Brasileiro de Paleontologia, VI Congreso Latino Americano de Paleontología. Resumenes. Sergipe, Brasil

Ubilla M, Perea D (1999) 6. Quaternary vertebrates of Uruguay. A biostratigraphic, biogeographic and climatic overview. In: Rabassa J, Salemme M (eds) Quaternary vertebrate paleontology in

South America, vol 12. Special volume of Quaternary of South America and Antarctic peninsula, pp 75–90

Vaz Ferreira R, De Zolessi C, Achaval F (1970) Oviposición y desarrollo de ofidios y lacertilios en hormigueros de Acromyrmex. Physis 29(79):431–459

Vaz Ferreira R, De Zolessi C, Achaval F (1973) Oviposición y desarrollo de ofidios y lacertilios en hormigueros de Acormyrmex. V Congreso Latinoamericano de Zoología. Actas 1:232–244. Montevideo, Uruguay

Verde M (2000) Trazas de Cerambícidos (Exapoda, Coleoptera) en madera fósil de edad Lujanense (Pleistoceno Superior) en el Uruguay. 1° Jorn de Paleontol del Uruguay. Resúmenes Ampliados: 30–32

Verde M, Ubilla M, Jiménez J, Genise J (2006) A new earthworm trace fossil from paleosols: aestivation chambers from the late pleistoce sopas formation of uruguay. Palaeogeogr Palaeoclimatol Palaeoecol 243:339–347

Verzi D (1998) Morfología craneo-dentaria de Abalosia castellanosi (Rodentia, Octodontidae, Octodontinae): filogenia, biogeografía y paleoambientes. VII Congreso Argentino de Paleontología y Bioestratigrafía, Resúmenes: 137. Bahía Blanca

Verzi D, Montalvo C (2008) The oldest South American Cricetidae (Rodentia) and Mustelidae (Carnivora): Late Miocene faunal turnover in central Argentina and the Great merican Biotic Interchange. Palaeogeogr Palaeoclimatol Palaeoecol 267(3–4):284–291

Verzi D, Quintana C (2005) The caviomorph rodents from the San Andrés Formation, east-central Argentina, and global Late Pliocene climatic change. Palaeogeogr Palaeoclimatol Palaeoecol 219(3–4):303–320

Verzi D, Tonni E, Scaglia O, San Cristóbal J (2002) The fossil record of the desert-Adapted South American rodent Tympanoctomys (Rodentia, Octodontidae). Paleoenvironmental and biogeographic significance. Palaeogeogr Palaeoclimatol Palaeoecol 179:149–158

Verzi D, Deschamps C, Tonni E (2004) Biostratigraphic and paleoclimatic meaning of the Middle Pleistocene South American rodent Ctenomys kraglievichi (Caviomorpha, Octodontidae). Palaeogeogr Palaeoclimatol Palaeoecol 212:315–329

Verzi D, Montalvo C, Deschamps C (2008) Biostratigraphy and biochronology of the Late Miocene of central Argentina: evidence from rodents and taphonomy. Geobios 41:145–155

Villwock J, Tomazelli L (1996) Holocene coastal evolution in Río Grande do sul Brazil. Quat S Am Antarctic Peninsula 12:283–296

Violante R, Parker G (2004) The post-glacial maximum transgression in the de la Plata River and adjacent inner continental shelf Argentina. Quatern Int 114:167–181

Vizcaíno S, Bargo M (1987) Los armadillos (Mammalia, Dasypodidae) del sitio arqueológico La Toma (Partido de Coronel Pringles, provincia de Buenos Aires). Aspectos paleoambientales relacionados. Terceras Jornadas Argentinas de Mastozoología, Resúmenes: 4. Buenos Aires

Voglino D (1999) Geología superficial y paleontología de las barrancas del Río Paraná entre Rosario (Santa Fé) y Campana (Bs. As.). Privately printed

Vucetich M (1986) Historia de los roedores y primates en Argentina: su aporte al conocimiento de los cambios ambientales durante el Cenozoico. IV Congreso Argentino de Paleontol y Bioestratigrafía, Actas 2:157–165

Vucetich M, Verzi D (1999) 12. Changes in diversity and distribution of the Caviomorph rodents during the Late Cenozoic in Southern South America. In: Rabassa J, Salemme M (eds) Quaternary vertebrate paleontology in South America. Special Volume of Quaternary of South America and Antarctic Peninsula, N°12, pp 207–223

Vucetich M, Verzi D (2002) First record of Dasyproctidae (Rodentia) in the Pleistocene of Argentina. Paleoclimatic implications. Palaeogeogr Palaeoclimatol Palaeoecol 178:67–73

Vucetich M, Verzi D, Tonni E (1997) Paleoclimatic implications of the presence of Clyomys (Rodentia, Echimyidae) in the Pleistocene of central Argentina. Palaeogeogr Palaeoclimatol Palaeoecol 128:207–214

Wilson EO, Holldobler B (1990). The Ants. p 732

Zárate M, Bargo M, Vizcaíno S, Dondas A, Scaglia O (1998) Estructuras biogénicas en el Cenozoico tardío de Mar del Plata (Argentina) atribuibles a grandes mamíferos. Asoc de Sedimentología 5(2):95–103

Zárate M, Schultz P, Blasi A, Heil C, King J, Hames W (2007) Geology and geochronology of type Chasicoan (late Miocene) mammal-bearing deposits of Buenos Aires (Argentina). J S Am Earth Sci 23:81–90

Zavala C, Navarro E (1991) Depósitos fluviales en la Formación Monte Hermoso (Plioceno Medio-Superior), provincia de Buenos Aires. XII Congreso Geol Argentino y II de Exploración de Hidrocarburos, Actas 2:236–244

Zavala C, Quattrocchio M (2001) Estratigrafía y evolución geológica del río Sauce Grande (Cuaternario), provincia de Buenos Aires, Argentina. Rev de la Asoc Geol Argent 56(1):25–37

Zucol A, Brea M, Scopel A (2005) First record of fossil wood and phytolith assemblages of the Late Pleistocene El Palmar Nacional Park (Argentina). J S Am Earth Sci 20:33–43

Zurita A, Scillato-Yané G, Carlini A (2005) Zoogeographic, biostratigraphic, and systematic aspects of the genus Sclerocalyptus Ameghino, 1891 (Xenarthra, Glyptodontidae) of Argentina. J S Am Earth Sci 20:121–129

Conclusions

Ichnology only recently incorporated studies on continental areas. This included an unusual accumulation of information that began to be elucidated and ordered. The sediments of the Late Cenozoic of the Pampasia offers innumerable and varied evidence of the development of the biota in paleosols and other stratigraphic units.

Vertebrate footprints were found in various locations and sedimentary outcrops of different ages: footprints of xenarthrans, macrauquenids, carnivorous marsupials, Homo, and forage birds were mentioned, as well as small and large caves of mammals.

Vertebrate and invertebrate coprolites were found in various localities in the Pampasia, from the Late Miocene to the Pleistocene. Coprolites contain traces of vertebrates (produced by predators) and plant elements, as well as traces of insects (produced by herbivores or omnivores).

Enteroliths or bezoars are elements produced or ingested by herbivore mammals. In the Late Pleistocene of Pampasia, a phosphate body associated with Scelidotherium remains is recorded.

The regurgitations come from the predatory activity of groups of birds, such as Strigidae, Tytonidae, Laridae and Ardeidae, inhabitants of ravines, caverns or forests, where these remains are concentrated. These accumulations are made up of remains of small mammals, fish, small birds and insects. Deposits of these materials are recorded at Pleistocene levels, as well Early Holocene deposits.

The crotovinas are traces of animal activity in continental lands; they represent the most numerous and varied icnites of the Late Cenozoic of the Pampasia. They correspond to the fossorial activity of numerous groups of vertebrates, especially mammals, varying in size and architecture. Some remains of its inhabitants were found in them (ctenomine and lagostomine rodents) in the Late Pliocene; in the Pleistocene the large caves attributed to xenartrans are significant, which left imprints of activity that allowed the identification of some taxa.

Rhizoliths or traces of roots are obvious signs of the presence of paleosols. They correspond to sedimentary organic structures that denote the ancient presence of plants and were ordered based on their morphology in several basic types: molds,

© The Editor(s) (if applicable) and The Author(s), under exclusive license to Springer Nature Switzerland AG 2020

J. H. Laza, *Ichnology of the Lowlands of South America*, Springer Earth System Sciences, https://doi.org/10.1007/978-3-030-62597-9

emptyings, tubules, root molds s.s. and petrifications. His study allows the inference about the prevailing climatic conditions in the past. The shape, as well as the association of the rhizoliths and their frequency can be excellent indicators of the conditions of plant development in the paleosols. Its presence is varied in numerous paleosols of the Pliocene and Pleistocene of the Pampasia.

The presence of crabs or crabs habitats responds to two types of environments: the strictly continental and the associated coastal marine areas that suffered advances in their waters during certain periods. In both cases, sub-aerial habits of their occupants are recognized. The ducts and chambers that make up their nests are recognized for being built by pellets. In the environments where they developed in the Pampasia two ichnoforms were recognized: (a) Psilonichnus, characterized by holes and vertical tubes, of supra coastal habits and (b) Ophiomorpha, characterized by the development of branched horizontal tubes, as inhabitants of shallow beaches. The former were found at levels corresponding to the "Belgranense" marine ingression, whereas the latter populated an extensive tidal zone in the southern sector of the Buenos Aires coast, as well as in sectors of the Uruguayan shore.

It has been relatively frequent to find fossil vertebrate remains that show signs of insect activity such as canaliculi and perforations, among others, generated by dermestid beetles and termites. Signs of activity attributable to these groups of insects were found in palaeoedaphic Pleistocene levels.

Various levels of paleosols are carriers of small size cylindrical tubes. Some of them have a smooth interior surface, with intersections and fecal remains, while others show menisci in their constitution. In the late Pleistocene, small spherical chambers similar to those of worm incubation were associated with these tubes. Different forms of these microtubules were found in paleosols from the late Miocene to the Holocene, belonging to periodically flooded levels; its construction is attributed to oligochaetes.

Mima type mounds are found in various sectors of the Pampean territory; they are elevations 2 m high and 10 m in diameter. Its presence has been attributed to two possible forms of origin: (a) rodent or dasipodid digging activity; (b) implantation of anthill. In both cases it is believed that animal activity developed over several generations.

The activity of different groups of insects, solitary or in community was recorded in numerous paleosols of the Pampasia. The development of ichnological studies allowed the creation of several ichnofamilies. The Ichnofamily Celliformidae groups the different structures, social or solitary of Hymenoptera and have been found in paleosols of the Miocene and Holocene of the region. The Ichnofamily Krausich-nidae comprises a varied group of ichnogenera that involves nests of termites and ants; its geographic distribution is wide and covers the entire Neogene. The Coprinis-phaeridae Ichnofamily corresponds to the beetle nests; it is one of the most common traces in the paleosols of the Cenozoic of South America. Most forms consist of a single reproduction chamber, although there are forms that accumulate several of them in a larger chamber. The vast majority of Coprinisphaeridae species are coprophagous, inhabiting plains that house an important herbivorous fauna; a few forms are microphages. The diverse genera and species identified in the numerous paleosols account for a varied and rich ichnobiota, which represents an enriching contribution to the understanding of the development of life at different times and geographic situations of the past.

Printed in the United States
by Baker & Taylor Publisher Services